ENDANGERED AND DISAPPEARING
BIRDS OF APPALACHIA
AND THE SOUTHEAST

ENDANGERED AND DISAPPEARING BIRDS OF APPALACHIA AND THE SOUTHEAST

Matt Williams

UNIVERSITY PRESS OF KENTUCKY

Copyright © 2024 by The University Press of Kentucky

Scholarly publisher for the Commonwealth, serving
Bellarmine University, Berea College, Centre College
of Kentucky, Eastern Kentucky University, The Filson
Historical Society, Georgetown College, Kentucky
Historical Society, Kentucky State University, Morehead
State University, Murray State University, Northern
Kentucky University, Spalding University, Transylvania
University, University of Kentucky, University of Louisville,
University of Pikeville, and Western Kentucky University.
All rights reserved.

Editorial and Sales Offices: The University Press of Kentucky
663 South Limestone Street, Lexington, Kentucky
40508-4008
www.kentuckypress.com

Map created with mapchart.net

Cataloging-in-Publication data available from the
Library of Congress

ISBN 978-0-8131-9836-1 (hardcover)
ISBN 978-0-8131-9896-5 (pdf)
ISBN 978-0-8131-9897-2 (epub)

This book is printed on acid-free paper meeting
the requirements of the American National Standard
for Permanence in Paper for Printed Library Materials.

Manufactured in the United States of America

 Member of the Association
of University Presses

CONTENTS

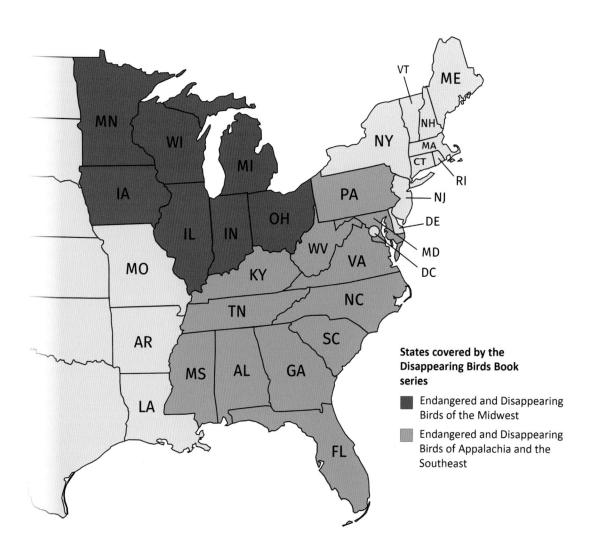

VT
ME
MN
WI
NH
MA
NY
CT
MI
RI
IA
PA
NJ
IL
IN
OH
DE
MO
WV
VA
MD
KY
DC
TN
NC
AR
SC
MS
AL
GA
LA
FL

States covered by the Disappearing Birds Book series

Endangered and Disappearing Birds of the Midwest

Endangered and Disappearing Birds of Appalachia and the Southeast

AUTHOR'S NOTE

The fifty species included in this text can be found in the twelve states shaded in green on the accompanying map—the area I am calling Appalachia and the Southeast. The birds covered in the following pages can be seen in this part of the country as they nest, winter, or migrate through the region at various times of the year. The seven states shaded in blue were covered in a previous book, *Endangered and Disappearing Birds of the Midwest.* Some declining species occur in both regions and therefore appear in both volumes.

Introduction
A Crisis of Disappearing Birds

According to recent estimates published in the journal *Science,* nearly 3 billion birds have been lost in the United States and Canada since 1970 (Rosenberg et al. 2019). Although some species have increased in number, these increases have been more than offset by significant decreases across a broad range of species and habitats. Taken together, the net result is a loss of 2.9 billion birds in the past fifty years. To put this into perspective, nearly 30 percent of the birds that once roamed the skies in this part of the world have disappeared within just a handful of decades.

What has caused such an alarming decline? A variety of factors are implicated, but loss of habitat is probably the single biggest contributor. The loss of wetlands, large stands of contiguous forests, grasslands, and other vital habitat has an impact on bird populations whether these habitat losses occur in breeding grounds, in migratory hot spots, or across Central and South America, where many of North America's birds spend the winter months. Although habitat loss is critical, other factors such as climate change, invasive species, pesticide use, light pollution over large cities, and even feral cats have likely played a role in the population decline of many bird species.

The purpose of this book is not to be a comprehensive field guide or a scientific reference for ornithologists. Other works accomplish these tasks admirably. Rather, this book tells the stories of some of the most beautiful and charismatic bird species across the Appalachians and the southeastern United States that are in serious trouble. In some cases, their populations are already perilously low; in other cases, their population trends are steeply negative, even if there are still a fair number of individual birds remaining. The birds included in this book appear on a variety of "watch lists" that identify species in need of conservation action (listed later). This book does not include every species in trouble; my hope is that by telling the stories of some of these incredible creatures and sharing compelling photographs of these birds in their natural habitats, even people with just a casual interest will be encouraged to help reverse the tide of disappearing birds.

By some estimates, bird-watching is one of the fastest growing hobbies in the United States. Birds are delightful, inspiring creatures with an amazing ability to navigate vast distances during migration; they can brighten our days and lift our moods with birdsong that fills the morning air or with their antics at the feeders outside our windows. Watching a nest full of eggs hatch and become balls of fluff that one day bravely test their wings is the type of experience that leads many to learn more about birds and how we can protect them. For some, that means keeping cats indoors or turning the lights out at night. For others, it means choosing shade-grown coffee or other products that reduce deforestation in the tropics. It can include improving the habitat in our own yards by growing native plant species, controlling populations of nonnative plants, or supporting land trusts and other organizations that protect bird habitat.

On many occasions I have wished that I could go back in time to witness a cloud of Passenger Pigeons flying overhead and darkening the sky for hours or to hear the calls of Carolina Parakeets—species that are now extinct. Recently, it occurred to me that in one hundred years, people may wish that they could travel back in time to our day and age to witness the incredible birds we presently enjoy. Without our help, the species we take for granted may not be with us just a few generations from now. I hope this book inspires readers to learn more about the incredible birds we have been blessed with and to take action now to ensure that these fantastic birds will still be around to inspire those who come after us.

In choosing the species to include in this work, the following sources were consulted: 2021 US Fish and Wildlife Service (USFWS) Birds of Conservation Concern, Breeding Bird Survey data, 2016 Partners in Flight Species of Continental Concern watch list, 2021 Audubon Priority Birds list, American Bird Conservancy watch list, US Shorebird Conservation Plan, 2016 State of North America's Birds report, and a variety of state lists that identify bird species of conservation concern. The estimates for population trends come from Breeding Bird Survey data where possible (1966–2019). For some species, however, the best trend estimates come from other sources, including aerial surveys, Christmas Count data, or counts performed at key migratory stopovers. In a few cases in which reliable trend information was not available, I provide a simple estimate of the total remaining population, based on numbers taken from a variety of sources.

When deciding which species to include, one of the most significant factors I considered was the species' Conservation Concern Score (CCS). This ranking mechanism was generated for all 1,154 native bird species in North America as part of the 2016 State of North America's Birds report. The overall CCS for each species is a combination of multiple factors, including current global population size, population trend, and threats to breeding and wintering habitat. Each factor is scored separately on a scale of 1 to 5, with 1 being the lowest concern and 5 being the highest concern. These scores are then added to come up with an overall score for each species ranging from 4 to 20, with 4 indicating the least conservation concern and 20 indicating the maximum concern. Birds that score in the 9 to 13 range are considered to be at the "moderate" concern level, while birds in the 14 to 20 range are in the "high" concern category. This book includes only species of at least moderate concern based on the CCS. To make my final determinations, I also considered the other conservation concern lists previously mentioned to arrive at the fifty species included in this book.

After reading about the incredible species highlighted in this book, you may find yourself wondering how you can support bird conservation and reverse the negative trends associated with so many of our bird species. Here are some actions you can take that will make a difference:

- Learn how to identify plants that are invasive and not native to your area. Invasive plants have no natural enemies and can spread incredibly quickly, altering the habitat and reducing the plant and insect foods available to birds. Plant only native species, and volunteer with local weed management groups to remove invasive plants and restore native ones.
- Encourage people to turn lights off at night, especially during the spring and fall migration. Light pollution can be disorienting to birds migrating at night, resulting in large numbers of birds dying when they strike towers, buildings, or other structures. This is especially true when it is foggy or visibility is reduced for other reasons. Hundreds of fatalities can occur at a single site in one night, and as many as 600 million birds die each year from collisions with buildings, communication towers, and power lines (Loss et al. 2014).
- Keep your cats indoors. Believe it or not, it is estimated that cats kill up to 4 billion birds every year (Loss, Will, and Marra 2013). Getting your cats spayed or neutered will help keep cat populations down, but just keeping them inside will eliminate the problem altogether. It is a simple solution that would make a huge difference for the birds.
- In the tropics, where many North American birds spend the winter months, forest habitat is being lost as it is converted for agricultural use, especially coffee plantations, which provide habitat for only a few bird species. Some plantations are experimenting with shade-grown coffee, leaving some of the larger canopy trees in place and growing the coffee plants underneath them. This type of plantation offers a more structurally diverse habitat that can attract and sustain a much broader group of birds. Buying shade-grown coffee brands can provide a market-based incentive for coffee growers to maintain better bird habitat. The Smithsonian has developed a "Bird Friendly" certification for coffee that is less harmful to the environment.
- Don't use insecticides. Although these chemicals can protect your garden, they have a negative impact on bird populations. Insect populations are declining at precipitous rates around the world, so it is no surprise that many birds that depend on insects for food are also declining rapidly. Not using insecticides on your property and choosing foods that are grown organically are great ways to promote a healthier environment for birds and many other species.
- Give financially or volunteer your time to conservation organizations that are protecting bird habitat. There are many great organizations working at the local, state,

national, and international levels to conserve habitat for birds and other life on earth. Supporting these organizations can have a huge impact on bird conservation.

There are many ways to make a difference and protect bird species. I hope this book inspires you to do just that. I am so grateful for the opportunity to enjoy the incredible, beautiful diversity of bird life that still thrives in many places across Appalachia and the Southeast. With hard work, we have witnessed many success stories about birds that have come back from the brink of extinction. Let's make sure these successes continue and that future generations have the opportunity to enjoy wild places and be thrilled by the beauty and majesty of life on earth.

Mottled Duck
(*Anas fulvigula*)

CONSERVATION CONCERN SCORE: 13 (Moderate)

OTHER DESIGNATIONS: American Bird Conservancy watch list (Red),
    State Protected (AL), Species of Greatest Conservation Need (FL, MS, SC)

ESTIMATED POPULATION TREND 1966–2019: –68%

SIZE: Length 22 inches; wingspan 30 inches

Mottled Ducks have a black mark at the base of the bill that
distinguishes them from the similar American Black Duck,
which lacks this marking. Male and female Mottled Ducks have
slightly different colored bills when seen in a good light.

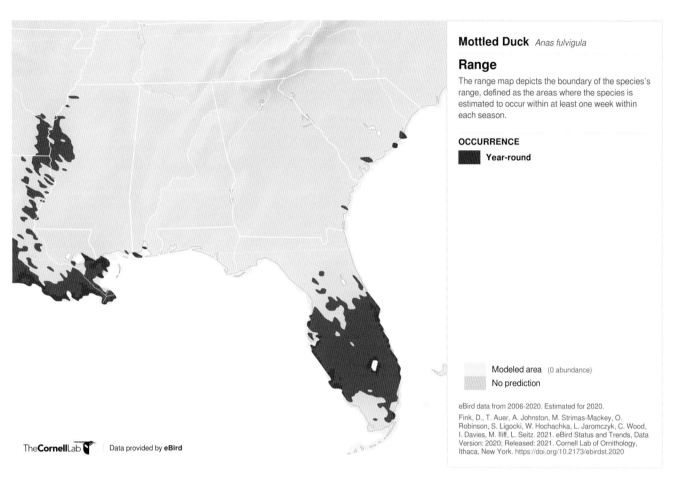

Mottled Duck *Anas fulvigula*

Range

The range map depicts the boundary of the species's range, defined as the areas where the species is estimated to occur within at least one week within each season.

OCCURRENCE

Year-round

Modeled area (0 abundance)

No prediction

eBird data from 2006-2020. Estimated for 2020.
Fink, D., T. Auer, A. Johnston, M. Strimas-Mackey, O. Robinson, S. Ligocki, W. Hochachka, L. Jaromczyk, C. Wood, I. Davies, M. Iliff, L. Seitz. 2021. eBird Status and Trends, Data Version: 2020; Released: 2021. Cornell Lab of Ornithology, Ithaca, New York. https://doi.org/10.2173/ebirdst.2020

TheCornellLab | Data provided by **eBird**

Species Account The Mottled Duck is unusual, in that it breeds in southern marshes rather than the prairie pothole region of the Dakotas and Canada, where most other dabbling duck species nest. Although they prefer the freshwater and brackish marshes and wetlands throughout peninsular Florida and along the western Gulf coast, Mottled Ducks have been known to use ditches, impoundments, and other human-made waterways in suburban environments as the naturally occurring wetlands have been drained or otherwise altered. In addition to their more southern nesting distribution, Mottled Ducks differ from other dabbling ducks in that they are typically found in pairs or small groups rather than large flocks. The exception occurs in prime feeding areas, where larger congregations are more common.

Mottled Ducks prefer shallow water surrounded by emergent vegetation for feeding. They filter substrates and surface water, searching for seeds or small invertebrates. They occasionally tip up to reach more deeply below the water's surface or reach up and use their bills to strip seeds from stems. Animal matter in the diet includes snails, crayfish, beetles, and the larvae and nymphs of several species. Occasionally, small fish are consumed. Favored seeds include cultivated rice, widgeon grass, and rushes.

Mottled Ducks are preyed on by a variety of predators. Ducklings are often taken by bullfrogs, snapping turtles, and large fish such as bass and gar; alligators feed on ducklings as well as adult birds. Northern Harriers and Peregrine Falcons are also known to prey on Mottled Ducks. A variety of mammals take the eggs or incubating hens, including skunks, raccoons, feral cats, and minks.

Mottled Ducks are seasonally monogamous, and pairs are formed as early as November. Their vocalizations and behaviors are quite similar to those of the closely related Mallard. There is occasional interbreeding between the two species, which produces a variety of hybrids humorously referred to as "Muddled Ducks." This mixing of species has been complicated by the growth of a domestic Mallard population consisting of released birds. The genetic mixing of domesticated Mallards and Mottled Ducks means that up to 10 percent of the Mottled Duck population now contains some Mallard DNA and is no longer pure. Wildlife officials have tried to educate the public about this problem and have even attempted to reduce the population of domesticated Mallards in some parts of Florida. If this trend continues, it could pose a serious threat to the Mottled Duck population, which has already been significantly reduced due to the loss of marsh and other wetland habitat.

Identification Mottled Ducks are a rich, warm brown and black throughout the back and belly. The plain buff-colored neck and head areas are noticeably free of streaking. Male Mottled Ducks have bright yellow bills with a black gape; females have an orange-colored bill with dark markings. The legs and feet are bright orange. In pure Mottled Ducks, the bluish-purple speculum has no white border above and only a thin, faint whitish border below. Mottled Ducks with Mallard DNA have stronger white borders above and below the speculum and more streaking on the neck and face.

Vocalizations Mottled Ducks give *quack* calls that sound very similar to the familiar calls of the Mallard.

Nesting The nest is constructed by the female in dense marsh or prairie vegetation, usually within six hundred feet of water. The nest is often hidden by overhanging grasses and is lined with grass and down feathers. Eight to twelve pale olive or dull white eggs are laid.

Mottled Ducks with some Mallard genes tend to have significant white bands above and below the blue speculum. This bird shows little or no white, indicating genetic purity.

Chimney Swift
(*Chaetura pelagica*)

CONSERVATION CONCERN SCORE: 12 (Moderate)

OTHER DESIGNATIONS: 2021 USFWS Birds of Conservation Concern,
Species of Greatest Conservation Need (FL, MD, PA, SC, TN, VA, WV)

ESTIMATED POPULATION TREND 1966–2019: –68%

SIZE: Length 5 inches; wingspan 14 inches

Chimney Swifts spend their waking hours in almost
constant flight. They are superb acrobats on the wing,
capable of making quick adjustments to capture insects.

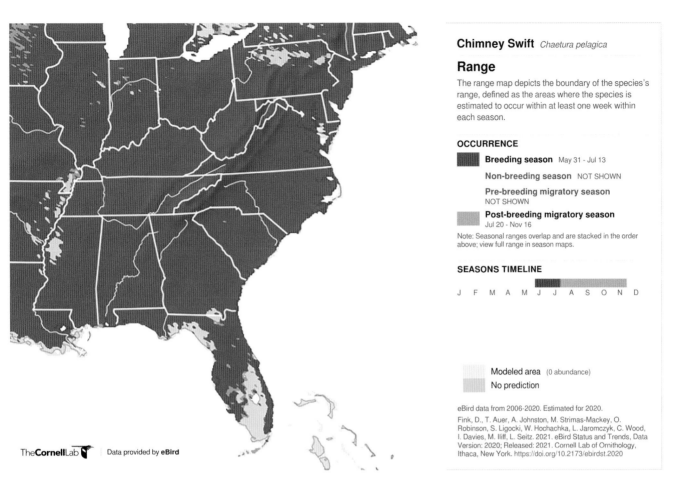

Chimney Swift *Chaetura pelagica*

Range

The range map depicts the boundary of the species's range, defined as the areas where the species is estimated to occur within at least one week within each season.

OCCURRENCE

Breeding season May 31 - Jul 13

Non-breeding season NOT SHOWN

Pre-breeding migratory season NOT SHOWN

Post-breeding migratory season Jul 20 - Nov 16

Note: Seasonal ranges overlap and are stacked in the order above; view full range in season maps.

SEASONS TIMELINE

J F M A M J J A S O N D

Modeled area (0 abundance)
No prediction

eBird data from 2006-2020. Estimated for 2020.
Fink, D., T. Auer, A. Johnston, M. Strimas-Mackey, O. Robinson, S. Ligocki, W. Hochachka, L. Jaromczyk, C. Wood, I. Davies, M. Iliff, L. Seitz. 2021. eBird Status and Trends, Data Version: 2020; Released: 2021. Cornell Lab of Ornithology, Ithaca, New York. https://doi.org/10.2173/ebirdst.2020

TheCornellLab | Data provided by **eBird**

Species Account Twittering high overhead on summer evenings, the distinctively shaped Chimney Swift can be seen over many cities and towns across the Appalachians and the Southeast. With stiff, rapid beats of their thin, curved wings, Chimney Swifts twist and turn at high speeds in their aerial pursuit of airborne insects. Other than when they are roosting at night or nesting, Chimney Swifts are almost perpetually on the wing—more so than just about any other land bird. They can bathe, gather nesting material, eat, drink, and even mate without landing. When they do land, they prefer rough vertical surfaces, where their stiff tails and long, sharp claws help hold them in place.

Shortly after returning to their breeding grounds, Chimney Swifts engage in aerial chases to establish pair-bonds. During these flights, three swifts may chase one another through the air, or two swifts may perform a V-wing display, simultaneously lifting their wings in a V and slowly drifting downward together. Once pairs are established and a nest site is chosen, swifts hover to pluck dead twigs from trees and then carry them to the nest site, where they use their saliva to attach the sticks to the inside of the chimney. Although many unmated swifts may use the same chimney to roost at night for protection from the elements and from predators, only one mated pair nests in each chimney.

During migration, many thousands of Chimney Swifts may roost together in a single large chimney. By adjusting how closely together they huddle, they can regulate their body temperature. On the coldest nights, swifts can enter torpor, a state of semihibernation that slows their metabolism and allows them to use less energy. In rare instances, swifts may huddle together on the outside of a large tree trunk if no other shelter is available.

Before human settlement, Chimney Swifts nested and roosted in natural structures such as caves and hollow trees. Once masonry chimneys became available, their population likely increased with this rapid expansion of suitable nesting sites. In recent years, however, the number of chimneys available for nesting has steadily decreased, as most modern chimneys are smaller and made of metal or plastic. Even the large brick chimneys that remain are often capped, and grates installed to keep pests out of chimneys keep Chimney Swifts out as well. Regular chimney maintenance also destroys many nests and young each year. Even so, recent studies indicate that there are still many suitable unoccupied chimneys, so there may be other factors driving the species' steep decline. For example, surveys designed to detect Chimney Swifts found that only 4 percent of suitable chimneys in North Carolina were occupied by swifts (Mordecai 2008). Collisions with towers kill a number of swifts every year, and in one case, more than one thousand Chimney Swifts were killed by vehicle collisions in a single day when cold weather kept insects flying low over a busy road (Bohlen 1989). However, these types of collisions are unlikely to affect overall population trends. A more likely cause is the decline of many insect species. Chimney Swifts eat large quantities of flying insects such as beetles, mayflies, wasps, and bees. The global decrease of these insect populations is likely a serious driver of declining populations of many insectivorous birds, including the Chimney Swift.

Identification The Chimney Swift is often referred to as a "cigar with wings" because of its long, cylindrical body. The birds are brownish gray overall, with thin, scythe-shaped wings and short, stubby tails.

Vocalizations The Chimney Swift's high-pitched, rapid, chittering call is heard most often in the evening as birds return to their roost sites in cities and towns after spending the daylight hours feeding, sometimes some distance away from the roost.

Nesting The nest is a stick platform built from small twigs attached to the wall of a chimney or hollow tree with the birds' saliva. There is some evidence that the saliva changes composition during nesting season, becoming stickier to better serve this purpose. Four or five white eggs are laid. Young birds may leave the nest at about twenty days of age, but they remain in the chimney for another week or so before venturing out for their first flight.

Courtship flights between male and female Chimney Swifts often involve close, synchronized displays that require incredible skill and are fascinating to watch.

King Rail
(*Rallus elegans*)

CONSERVATION CONCERN SCORE: 15 (High)

OTHER DESIGNATIONS: American Bird Conservancy watch list (Yellow),
2021 USFWS Birds of Conservation Concern, 2021 Audubon Priority Birds list,
In Need of Management (TN), State Special Concern (GA), Species of Greatest
Conservation Need (AL, FL, GA, KY, MD, MS, NC, PA, SC, VA)

ESTIMATED POPULATION TREND 1966–2019: −88%

SIZE: Length 15 inches; wingspan 20 inches

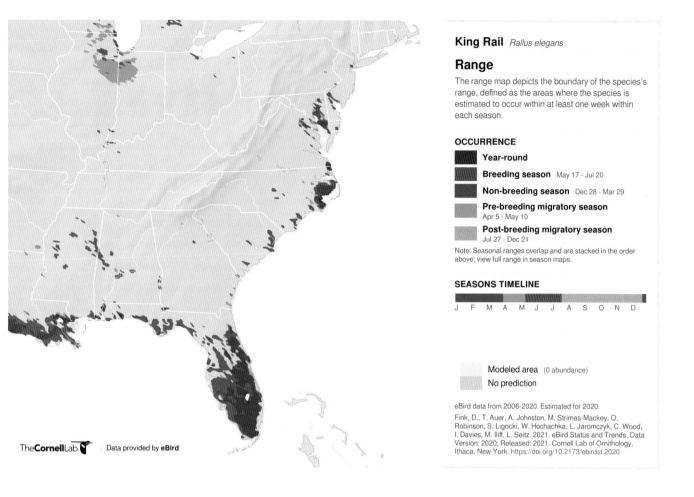

King Rail *Rallus elegans*

Range

The range map depicts the boundary of the species's range, defined as the areas where the species is estimated to occur within at least one week within each season.

OCCURRENCE

Year-round

Breeding season   May 17 - Jul 20

Non-breeding season   Dec 28 - Mar 29

Pre-breeding migratory season
Apr 5 - May 10

Post-breeding migratory season
Jul 27 - Dec 21

Note: Seasonal ranges overlap and are stacked in the order above; view full range in season maps.

SEASONS TIMELINE

J F M A M J J A S O N D

Modeled area   (0 abundance)
No prediction

eBird data from 2006-2020. Estimated for 2020.
Fink, D., T. Auer, A. Johnston, M. Strimas-Mackey, O. Robinson, S. Ligocki, W. Hochachka, L. Jaromczyk, C. Wood, I. Davies, M. Iliff, L. Seitz. 2021. eBird Status and Trends, Data Version: 2020; Released: 2021. Cornell Lab of Ornithology, Ithaca, New York. https://doi.org/10.2173/ebirdst.2020

TheCornellLab | Data provided by **eBird**

Species Account Once a common bird in freshwater and brackish marshes throughout the Southeast and the Midwest, the King Rail has suffered a severe population decline as wetlands have been destroyed across its range to make way for agriculture and development. Even where wetlands remain, they are often so altered by flood-control structures or invasive species such as reed canary grass or phragmites that they no longer provide suitable habitat for this species. Today, the best places to find populations of the King Rail are near the Gulf coast in Louisiana and Florida and along the Atlantic coast in the Carolinas. In these locations, its range overlaps with that of the very similar Clapper Rail, which prefers salt marsh rather than freshwater. Key habitat for King Rails often includes plant species associated with shallow water levels, including cattails, bulrushes, and spartina grasses.

King Rails exhibit more rich coloration on their cheeks than the similar Clapper Rail, which usually has gray cheeks.

Although the King Rail is a secretive bird that prefers to remain hidden in dense vegetation deep in impenetrable marsh, it sometimes rewards patient observers as it stalks along the water's edge in the open. Typically, however, these birds are most easily identified by their loud vocalizations during nesting season, so learning their calls is the best way to find them. In fact, these loud calls led to one of the bird's many vernacular names, "Stage Driver," because the *check-check* call reminded early farmers of how stagecoach drivers "talked" to their horses. Shortly after nesting season, King Rails undergo a complete molt of their feathers and are flightless for almost a month. During this time, they are especially secretive, virtually silent, and extremely difficult to find.

King Rails feed heavily on animals during much of the year, consuming more than 90 percent animal matter in the spring and summer. They switch to more plant matter during the winter months, consisting of mostly fruits and seeds. Crayfish and crabs make up much of their diet, but they eat a wide variety of prey items, including grasshoppers, fish, beetles, frogs, spiders, dragonflies, clams, and, rarely, small mammals. King Rails often eat smaller prey whole, but they may take larger items to a high point, such as a muskrat lodge, where they can tear the food into smaller pieces. The birds often dunk their prey items in water before consuming them, even if they have been captured some distance from the water. Indigestible bits of shells and exoskeletons are regurgitated as pellets.

It is estimated that the total global population of the King Rail stands at only about seventy thousand birds today. Although habitat loss is the main driver of the decline, other causes have played a role and will continue to do so. King Rails migrate at night and often strike towers, wires, and buildings. Many birds are inadvertently ensnared and killed in muskrat traps placed in wetlands, and others are threatened by lead poisoning, farm chemical runoff, and other pollutants. Despite the rapidly declining population, several states in the South still permit a hunting season for King Rails, and small numbers of birds are taken each year.

Identification The King Rail is the largest North American rail species. Although very similar to the Clapper Rail, the King Rail is generally more richly colored, with a more striking barred pattern of black and white on the flanks. The King Rail has a rich, rusty-orange breast and a more subdued pattern on its back consisting of feathers with a black center and buffy-orange edges. Its bill is long and narrow and slightly downcurved, and its legs are olive to light brown. King Rails often walk with their stubby tails pointed up, revealing snow-white undertail coverts.

Vocalizations King Rails give a variety of calls that can easily be confused with those of the Clapper Rail. The most common call is a long series of short *kek* or *check* sounds that are slightly lower, louder, and more consistent in tempo than the call of the Clapper

Rail, which often accelerates. Other vocalizations include a grating *k-kerrr* and a variety of softer calls made by adult birds when communicating with chicks.

Nesting The male often feeds the female during courtship and is heavily involved in nest building. The nest is placed within a clump of grasses or sedges. It consists of a raised platform with a depression in the middle that is shaped as the bird sits and rotates its body. Vegetation is bent over the nest to form a dome. There is usually some type of ramp leading down from the nest platform. Ten to twelve eggs are laid. The eggs are pale buff with light-brown speckles.

King Rails have more distinct black-and-white barring on the flanks than Clapper Rails. King Rails often lift their tails as they walk through the marsh, revealing the bright white underside. The male also flashes his bright white tail coverts during courtship.

Clapper Rail
(*Rallus crepitans*)

CONSERVATION CONCERN SCORE: 13 (Moderate)

OTHER DESIGNATIONS: American Bird Conservancy watch list (Yellow),
  2021 Audubon Priority Birds list, Species of Greatest Conservation Need (NC, SC, VA)

ESTIMATED POPULATION TREND 1966–2019: −39%

SIZE: Length 15 inches; wingspan 19 inches

The Clapper Rail is more grayish overall than the King Rail,
especially on the face. Its flanks are gray and white rather
than the distinct black and white of the King Rail.

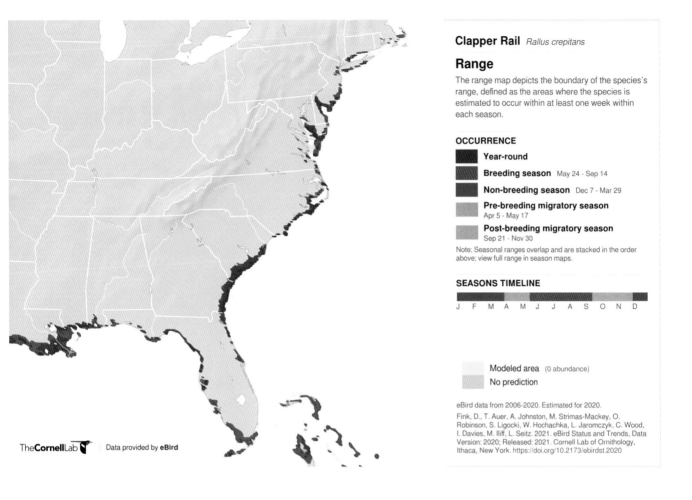

Clapper Rail *Rallus crepitans*

Range

The range map depicts the boundary of the species's range, defined as the areas where the species is estimated to occur within at least one week within each season.

OCCURRENCE

Year-round

Breeding season May 24 - Sep 14

Non-breeding season Dec 7 - Mar 29

Pre-breeding migratory season
Apr 5 - May 17

Post-breeding migratory season
Sep 21 - Nov 30

Note: Seasonal ranges overlap and are stacked in the order above; view full range in season maps.

SEASONS TIMELINE

J F M A M J J A S O N D

Modeled area (0 abundance)
No prediction

eBird data from 2006-2020. Estimated for 2020.

Fink, D., T. Auer, A. Johnston, M. Strimas-Mackey, O. Robinson, S. Ligocki, W. Hochachka, L. Jaromczyk, C. Wood, I. Davies, M. Iliff, L. Seitz. 2021. eBird Status and Trends, Data Version: 2020; Released: 2021. Cornell Lab of Ornithology, Ithaca, New York. https://doi.org/10.2173/ebirdst.2020

TheCornellLab | Data provided by **eBird**

Species Account The Clapper Rail was once so numerous along the Atlantic coast that renowned artist and ornithologist John James Audubon noted that it was possible to find one hundred nests of this species in a single day (Kaufman n.d.). The birds were very popular as a food source in the late 1800s, and many birds and eggs were taken for food, sport, or profit. A two-day hunt in New Jersey in 1896 produced a total take of tens of thousands of birds. Even though numbers had already started to decline, the annual reported harvest in Virginia was twenty-five thousand birds in 1922. The Clapper Rail remains a hunted species today in nearly every East Coast state, but harvest numbers are much lower, reflecting both fewer hunters and a reduced Clapper Rail population. More recent harvest numbers from Maryland, for example, reflect an annual harvest of only two hundred birds for the entire season. Although the Breeding Bird Survey (BBS) indicates a decline in this species, these data alone are insufficient to understand population trends, as few BBS routes encounter Clapper Rails. It is clear that populations have declined dramatically since the early 1900s, but more recent population trends are hard to project because this species is difficult to study. However, it is likely that Clapper Rail populations are continuing to decline as salt marsh habitat is lost to development, the hydrology is altered, or the habitat is degraded by invasive species.

There is considerable geographic variation within this species, and as many as eight subspecies have been described. Although some authors once considered the Clapper Rail and the King Rail to be the same species, each of them has now been split into additional species. For example, Clapper Rail populations in the southwestern United States and along the Pacific coast of Mexico are now known as Ridgeway's Rail, and Clapper Rail populations in Central and South America are now referred to as the Mangrove Rail.

Named for its loud, clapping call, the Clapper Rail can still be found in salt marshes from Massachusetts to the Gulf coast of northern Mexico. Clapper Rails are more easily heard than seen, although the patient observer may be rewarded with glimpses of the birds as they patrol mudflats and tidal channels in search of crustaceans. Fiddler crabs make up the bulk of their diet, although other crabs, clams, worms, and small fish are also taken. In the winter, Clapper Rails eat some seeds and other plant matter. But even at this time of year, vegetable matter makes up only about 10 percent of their diet, according to a study from Georgia (Oney 1951).

Avian predators take a toll on the species, with the Red-tailed Hawk, Northern Harrier, Great Horned Owl, Blue Heron, and Short-eared Owl all reportedly taking adult Clapper Rails. When Clapper Rails spot an avian predator, they give a sharp alarm call and seek heavy vegetation. In some cases, they may actually dive underwater to avoid an aerial attack. Raccoons, opossums, coyotes, and minks all likely take eggs, young, and adult birds. However, the most frequent cause of nest loss is storms that flood nests constructed just above the normal high-tide line. The Clapper Rail has been known to renest as many as five times in response to lost clutches. This resiliency, along with a large clutch size, helps make up for these frequent losses.

Identification The Clapper Rail is a large chicken-sized bird with long legs and a long neck. It is muted gray overall and has a brown back with black feather centers. The flanks are heavily barred with black and white, and the bill is reddish orange at the base shading to black at the tip. The Clapper Rail is more grayish, especially on the cheeks, than the closely related King Rail, which tends to have warmer reddish tones. Clapper Rails are seldom seen outside of salt marsh habitats, whereas King Rails prefer freshwater marshes; however, there is some overlap in brackish waters.

Vocalizations The Clapper Rail gives a series of loud, fast, short clapping calls that slow in pace toward the end. The birds also have a variety of grunting and growling calls that can carry some distance through the marsh.

Nesting Studies from Georgia and Mississippi indicate that Clapper Rails prefer to nest near or on the ground close to the high-tide line in dense vegetation, such as cordgrass clumps, and in proximity to either a tidal creek or tidal pool (Rush et al. 2010). Eighty percent of nests studied in Virginia were within fifteen feet of a tidal slough (Kozicky and Schmidt 1949). A ramp built of plant material often leads up to the level of the nest, with a canopy of vegetation woven over the nest. Seven to eleven eggs are laid. The eggs are pale yellow or olive, with brown blotches.

Clapper Rails are at home in the salt marshes of the Atlantic and Gulf coasts. They are most easily located by voice, but a patient observer can catch glimpses of the birds in spartina grasses or tidal sloughs.

Purple Gallinule
(*Porphyrio martinica*)

CONSERVATION CONCERN SCORE: 12 (Moderate)

OTHER DESIGNATIONS: Species of Greatest Conservation Need (FL, MS, SC, TN)

ESTIMATED POPULATION TREND 1966–2019: −19%

SIZE: Length 13 inches; wingspan 22 inches

Purple Gallinules use their long toes to help distribute their weight, which allows them to balance (sometimes precariously) on floating aquatic vegetation. If they start to sink, they quickly clamber to nearby vegetation or flap their wings to make themselves lighter.

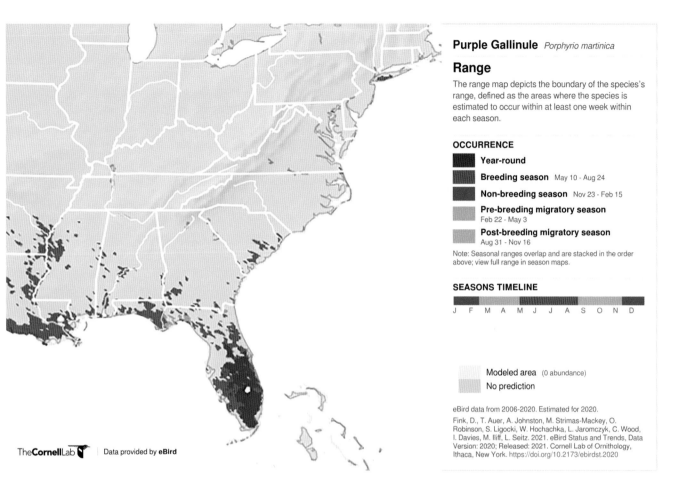

Purple Gallinule *Porphyrio martinica*

Range

The range map depicts the boundary of the species's range, defined as the areas where the species is estimated to occur within at least one week within each season.

OCCURRENCE

	Year-round
	Breeding season May 10 - Aug 24
	Non-breeding season Nov 23 - Feb 15
	Pre-breeding migratory season Feb 22 - May 3
	Post-breeding migratory season Aug 31 - Nov 16

Note: Seasonal ranges overlap and are stacked in the order above; view full range in season maps.

SEASONS TIMELINE

J F M A M J J A S O N D

	Modeled area (0 abundance)
	No prediction

eBird data from 2006-2020. Estimated for 2020.

Fink, D., T. Auer, A. Johnston, M. Strimas-Mackey, O. Robinson, S. Ligocki, W. Hochachka, L. Jaromczyk, C. Wood, I. Davies, M. Iliff, L. Seitz. 2021. eBird Status and Trends, Data Version: 2020; Released: 2021. Cornell Lab of Ornithology, Ithaca, New York. https://doi.org/10.2173/ebirdst.2020

TheCornellLab | Data provided by **eBird**

Species Account Sporting colors that include metallic purple, iridescent green, fire-engine red, sky blue, and brilliant yellow, the Purple Gallinule is undoubtedly one of the most beautiful and unusual birds in all of North America. Known in some areas as the "Lily Trotter," these long-legged birds can often be seen walking on lily pads using their long yellow toes to help distribute their weight. They can sometimes be as awkward and uncoordinated as they are beautiful, occasionally throwing their wings out to maintain their balance or flapping briefly if they prove too heavy for the floating vegetation they are walking on.

Unlike the closely related Common Gallinule that feeds as it swims, the Purple Gallinule uses its beak to flip over vegetation to look for aquatic insects, or it uses its long legs and toes to pull down tall stalks and stand on them to bring seeds into easier eating range. Generally, their diet is more vegetable than animal matter, but the birds take advantage of a wide variety of food sources. Common food items include flowers and fruits of the water lily family, as well as seeds, flowers, and tubers from a variety of other species, including pickerelweed, smartweed, and sedges. Beetles, spiders,

and dragonflies are also taken. When a larger prey item like a frog is captured, it often results in a flurry of activity, as other Purple Gallinules rush in and try to steal part of the meal. Occasionally, eggs or young are taken from other birds' nests, including the Common Grackle, Snowy Egret, Anhinga, and Wood Duck.

In areas where rice farms are found, rice is a major food source, and Purple Gallinules nest in rice fields once the plants reach a sufficient height. In a South American study, the stomach contents of birds nesting in rice fields were found to be 68 percent rice grains on average (McKay 1981). This has led some to consider the Purple Gallinule an agricultural pest species, although it is doubtful that these birds have much impact on overall productivity levels. In some parts of the world, Purple Gallinule served on a bed of rice is considered a delicacy. Although they are legal game birds in much of the United States, harvest numbers are quite low because most of the birds move farther south before hunting seasons begin.

Although the Purple Gallinule usually makes only short, weak flights in the marshes where it lives, the species is capable of long-range migration and has appeared in many northern states outside its normal range. In addition, these birds have been recorded as far away as Iceland, Switzerland, and parts of Africa. It appears that many of these vagrant birds were responding to extended drought conditions that triggered them to seek habitat elsewhere.

Across North America, populations of Purple Gallinules have declined as freshwater marshes have been drained. In some states in the Southeast, the decline has been especially pronounced. Populations in Florida have dropped by 56 percent, and the population in Alabama has declined by as much as 72 percent. Despite these habitat losses, other habitat has been added. Purple Gallinules use marsh habitat in cities and towns associated with human-made ponds and lakes created in local parks. This species may continue to decline as new

Purple Gallinules often use their long toes to flip over large leaves or pull down aquatic vegetation as they search for food.

varieties of rice that can be harvested more quickly are introduced, as the birds may not have enough time to rear their young in these habitats.

Identification Adults are brilliant metallic purple on the head, neck, and chest, with green and bronze colors on the back. Bright white feathers under the tail are often visible, as the birds tend to lift their tails as they walk. The forehead has a sky-blue shield above the red and yellow beak, and the legs are brilliant yellow. Young birds are a dull buff and cream on the head, neck, and chest, with some areas of olive and blue on the back and wings.

Vocalizations The birds give a variety of cackling calls that include some grunting, squawking, and rapid chicken-like clucking. They also make a sharper *kik* call. The birds are quite vocal during the breeding season, especially early in the morning.

Nesting The nest, consisting of a platform of grass, cattail, or other vegetation within a few feet of the water level, is constructed by both the male and the female. The female lays six to eight buff-colored eggs that are usually speckled with brown. The chicks are capable of walking as soon as they are hatched but rely on the parents to bring them food. The young birds are capable of flight at nine weeks.

Although most often seen in freshwater marshes, Purple Gallinules occasionally wander into drier habitats to search for insects.

Limpkin
(*Aramus guarauna*)

CONSERVATION CONCERN SCORE: 11 (Moderate)

OTHER DESIGNATIONS: State Special Concern (FL, GA)

ESTIMATED POPULATION TREND 1966–2019: Not available

SIZE: Length 26 inches; wingspan 40 inches

Limpkins' heavy, crane-like bills are specifically designed
to match the curve of the apple snail shell, allowing
them to access the snail without breaking the shell.

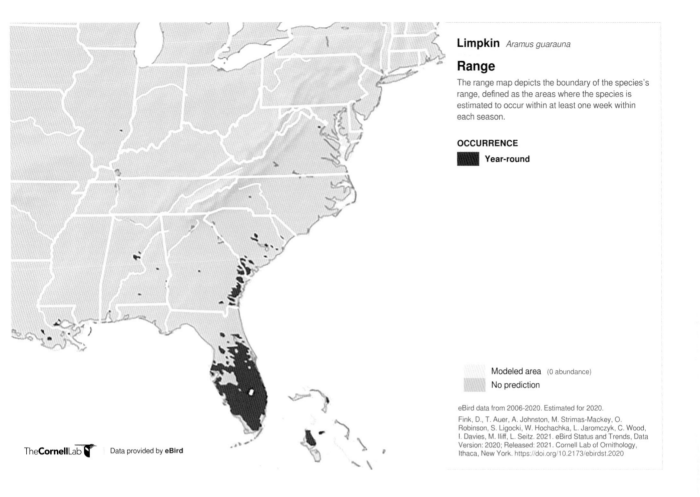

Species Account Like an enormous brown rail, the Limpkin inhabits freshwater wetlands throughout much of peninsular Florida, as well as in the Caribbean, Central America, and portions of South America. Dietary specialists, Limpkins in Florida eat apple snails almost exclusively. These birds have a uniquely shaped beak that curves downward and angles slightly to the right at the tip, which allows them to perfectly match the right-hand curve of the apple snail's shell. The tip of the beak is used almost like tweezers to remove the snail without breaking the shell. Limpkins often use the same feeding areas repeatedly, leaving piles of unbroken apple snail shells as a clue to the birds' presence. Based on the shells left at one such site in Florida, the birds' diet consisted of 76 percent apple snails, 13 percent freshwater mussels, and 11 percent other snail species (Bryan 1981).

Limpkins can forage by sight while wading in shallow water or by tactile sensation while sweeping their beaks along the marsh substrate. This means they are capable of successfully foraging at all hours of the day and night. The birds sometimes walk on floating vegetation such as water hyacinth or pickerelweed and probe it with their beaks in search of snails clinging to the underside of the mat. Despite their strong preference for snails under normal conditions, Limpkins have reportedly captured grasshoppers and other insects during times of drought.

Limpkins are quite territorial during the nesting season. Males defend their territories by charging intruders and chasing rivals through the marsh and even up angled tree trunks and low-hanging limbs. Males also engage in raucous calling displays that neighboring males often respond to—sometimes engaging a chain of males for a fair distance where numerous territories are laid out in linear succession. Male birds have unusually shaped primary feathers on the outer edge of the wing that can create a buzzing noise known as winnowing, which is used in flight displays to mark their territory.

It is thought that the name Limpkin originated with early settlers of the Florida peninsula because the birds appeared to walk with a limping gait, although this behavior has not been described in the literature. In fact, Limpkins are capable of impressive speed on the ground and seem quite agile. Hunted extensively by early settlers, the birds were apparently very tame and easy to shoot. They were considered excellent table fare and were hunted nearly to extinction in many parts of Florida by 1900. Several large-scale wetland drainage projects to increase agricultural lands further decimated the population in the northern Everglades, the Kissimmee River Valley, and St. John's Marsh, which reportedly had the largest breeding population of Limpkins in eastern Florida and may have boasted five hundred breeding pairs until it was drained and farmed in the early 1950s. Very little information exists on current population trends in Florida, and almost no information is available throughout the remainder of the Limpkin's range. However, given the loss of wetlands in Florida, the influx of nutrients from agriculture and the resulting impact on water quality, and the effects of invasive aquatic species, it is likely that the US population continues to decline. A 1994 survey estimated the total US Limpkin population to be three thousand to six thousand pairs.

Identification Limpkins in the United States are brown with white spots on the back and extensive white streaking on the head and neck. In parts of South America, the birds' backs are brown without the white spots. They have long grayish legs and toes that allow them to walk on vegetation or wade in shallow water. Their long, curved beaks are pale orange with a black tip.

Vocalizations Male birds have a long, looped trachea similar to cranes, which increases the carrying distance of their piercing *kreeow* calls. This call may be given in flight or during territorial disputes.

Nesting Nests are placed in a surprising variety of locations, such as in floating vegetation just above the water, in grasses, or among bald cypress knees. Nests have also been recorded higher up in trees, in old osprey nests, or even in tree cavities. The species may nest in loose colonies when food is readily available. Four to eight light-brown eggs with gray splotches are typically laid.

Limpkins in South America have plain brown backs that lack the bright silvery spots of the Florida population.

Mississippi Sandhill Crane
(*Antigone canadensis pulla*)

CONSERVATION CONCERN SCORE: Not available. Although the global population of Sandhill Cranes is secure and has a CCS of 9, the Mississippi Sandhill Crane would score much higher.

OTHER DESIGNATIONS: State Protected (AL), State Endangered (MS), Federally Endangered

ESTIMATED POPULATION TREND 1966–2019: Not available

SIZE: Length 42 inches; wingspan 84 inches

Because the range of the Mississippi Sandhill Crane is so small, it is not feasible to include a map. Generally, this subspecies is found only in extreme southeastern Mississippi in the immediate vicinity of the Mississippi Sandhill Crane National Wildlife Refuge.

Species Account The Mississippi Sandhill Crane, a nonmigratory subspecies of the Sandhill Crane, once occurred across much of the central Gulf coast from Louisiana east through Mississippi, Alabama, and the Florida Panhandle. Smaller, shorter-winged, and darker than the Sandhill Cranes found in much of eastern North America, the Mississippi Sandhill Crane is one of six subspecies that occur on the North American continent and in the Caribbean. Three of these subspecies are protected because of their declining or critically small populations: the Florida Sandhill Crane (state threatened), the endangered Cuban Sandhill Crane, and the Mississippi Sandhill Crane, first listed as federally endangered in 1973. The entire population of this subspecies is found on or near the Mississippi Sandhill Crane National Wildlife Refuge in coastal Mississippi near Gautier. The Nature Conservancy purchased the original lands for the refuge, and in 1975 the US Fish and Wildlife Service created the refuge for the specific purpose of recovering this population of cranes.

Mississippi Sandhill Cranes are shorter and darker than other Sandhill Cranes and need wet pine savanna habitat to survive. This bird has been fitted with color bands and a transmitter to help scientists track the recovery of these rare birds.

Young Mississippi Sandhill Cranes typically take three to four years to find a mate and attempt to nest. After a successful pair is formed, it may take a few more years before they lay viable eggs. Even experienced pairs rarely raise more than one chick, or colt, per year. Predators take a toll on nesting success, with bobcats, armadillos, and raccoons commonly preying on crane eggs. All these factors mean that recovery of the species is slow and requires a long-term effort. To increase reproductive success and augment the population, a captive breeding program has been established that releases ten to fifteen young birds each year. The Freeport-McMoRan Audubon Species Survival Center and White Oak Conservation have both been involved in rearing cranes for release in the wild. These programs have been quite successful, and upward of 90 percent of the free-flying Mississippi Sandhill Cranes in the wild today are captive-reared birds.

Mississippi Sandhill Cranes are omnivores and consume a wide variety of animal and plant matter. During the summer, they feed heavily in wet pine savanna habitats and capture a variety of insects, crayfish, frogs, earthworms, and small reptiles. At other times of the year, nuts, tubers, seeds, and fruits make up a greater percentage of the diet. Waste grain is consumed in the fall months, especially corn. Pecans are favored by the birds from September through December, and they often visit pecan orchards to feed. During late winter and early spring, courtship activity and pair-bond strengthening take place, even among mated pairs.

Following World War II, much of the native wet pine savanna of the central Gulf coast region was converted to rows of pine plantations as part of an effort to maximize timber production. Wetlands were also drained at this time, and naturally occurring fires that helped maintain the native pine savannas were more effectively controlled. In addition, a number of new roads and highways were constructed that further fragmented the cranes' remaining habitat. As the habitat was destroyed, the crane population declined. The last breeding records are from the early 1900s for Louisiana and from 1960 for Alabama. The total number of wild Mississippi Sandhill Cranes was estimated to be fifty to sixty birds in 1969. Numbers continued to decline and reached a low of twenty-five to thirty-five birds by 1986. Since that time, however, the population has steadily grown, thanks to the work and dedication of organizations focused on protecting and managing the remaining habitat and increasing the birds' reproductive success through captive breeding programs. As of March 2019, the wild population of this subspecies was estimated to stand at 129 individuals, including forty pairs.

Identification The Mississippi Sandhill Crane is shorter and darker than other Sandhill Cranes, making the white patch on the cheek look brighter. The birds are gray overall, with long, dark legs. They have bright yellow eyes and attain a red patch of skin on the forehead as they mature, appearing after their first year.

Vocalizations Calls are a loud, trumpeting *karrrroo* that can carry a great distance.

Nesting Nests are placed on the ground, often in a clump of grasses or vegetation near standing water. Typically, one clutch of one or two eggs is laid per year. Eggs are pale buff with brown markings. Newly hatched colts are yellowish and capable of leaving the nest and following their parents shortly after hatching.

Whooping Crane
(*Grus americana*)

CONSERVATION CONCERN SCORE: 16 (High)

OTHER DESIGNATIONS: American Bird Conservancy watch list (Red),
    State Protected (AL), Species of Greatest Conservation Need (FL, GA, KY, TN),
    Federally Endangered

ESTIMATED POPULATION TREND 1966–2019: Not available

SIZE: Length 54 inches; wingspan 87 inches

An adult Whooping Crane gives a loud, trumpeting call in response to its mate.

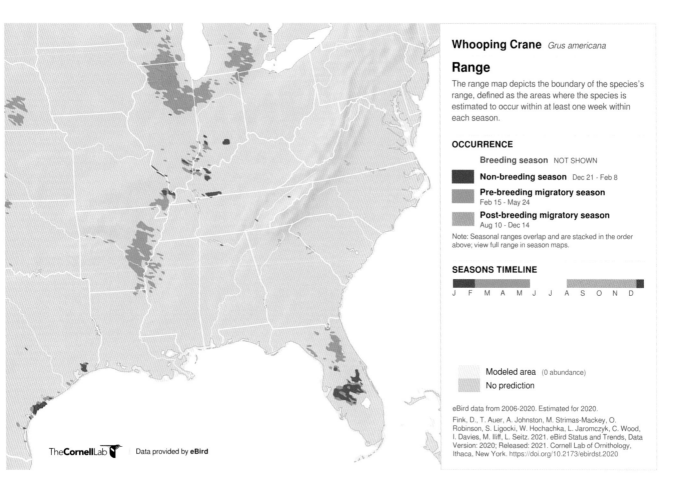

Species Account Standing nearly five feet tall, weighing fifteen pounds, and boasting a wingspan of more than seven feet, Whooping Cranes are truly magnificent creatures. The tallest bird in North America, the Whooping Crane has a resounding, bugling call that can be heard for more than two miles. During migration, the calls of these birds can be heard from great heights by observers on the ground.

Whooping Cranes' pair-bonds are created and strengthened through a variety of behaviors, including stylized walking and calling. For example, in the unison call, one bird begins the call and is joined almost simultaneously by its mate. Elaborate dances also help strengthen the bond. In his 1952 monograph titled *The Whooping Crane*, Robert Porter Allen describes the dance of a mated pair of cranes:

> Suddenly one bird (the male?) began bowing his head and flapping his wings. At the same time he leaped stiffly into the air, an amazing bounce on stiffened legs that carried him nearly three feet off the ground. In the air he threw his head back so that the bill pointed skyward, neck arched over his back. Throughout this leap the great wings were constantly flapping, their long black flight feathers in striking

contrast to the dazzling white of the rest of the plumage. The second bird (the female?) was facing the first individual when he reached the ground after completing the initial bounce. This second bird ran forward a few steps, pumping her head up and down and flapping her wings. Then both birds leaped into the air, wings flapping, necks doubled up over their backs, legs thrust downward stiffly. Again they leaped, bouncing as if on pogo sticks. On the ground they ran towards each other, bowing and spreading their huge wings.

During these great dancing displays, the birds often pick up small sticks or other loose material and toss it into the air. Pair-bonds typically last for the birds' lifetime. The attachment is so strong that if one bird is injured and unable to migrate to the breeding grounds, its mate often stays behind rather than migrate alone.

Although there is some debate about how numerous these birds once were, there may have been more than ten thousand individuals at one time. By the 1870s, the population had already declined to less than fifteen hundred birds. The major concentration of birds nested in the northern prairies of the United States and Canada and wintered along the Gulf coast and in parts of Mexico. This mirrors the path of the only remaining self-sustaining population of Whooping Cranes today, which breeds in the Wood Buffalo National Park (WBNP) in northwestern Canada and winters at Aransas National Wildlife Refuge in Texas. Records of Whooping Cranes wintering along the Atlantic coast from New Jersey to the Carolinas and Florida persisted into the 1900s. Some have speculated that this was a second major flock that nested somewhere near Hudson Bay and migrated to the southeastern United States. A third population was nonmigratory and lived year-round in coastal Louisiana.

By 1941, there were only twenty-one birds in two small populations—the flock that wintered at Aransas and had breeding grounds in Canada (the precise location at WBNP was unknown until 1954), and the nonmigratory population in Louisiana. The Louisiana population rapidly declined, and by 1950, only one bird remained. It was translocated to Aransas but did not survive the summer.

Faced with the likelihood of the species' extinction, biologists across Canada and the United States collaborated to develop innovative measures to save the Whooping Crane. Beginning in 1967, eggs from the wild at WBNP were removed from nests to establish a captive breeding program. This program has been tremendously successful and has provided the birds used in an attempt to create an additional self-sustaining wild population as part of the recovery plan. Unfortunately, efforts to do so in Idaho and Florida have been unsuccessful. A more aggressive effort involved rearing birds in Wisconsin and using ultralight aircraft to lead them to wintering grounds in Florida. Although this population was able to migrate successfully, further augmentation efforts have been discontinued due to issues related to reproduction within this flock. In 2010 a fourth effort was initiated to build a nonmigratory population in formerly occupied habitat in Louisiana.

In flight, Whooping Cranes are entirely white, except for large black primary feathers near the wing tips.

Today, the total wild population of Whooping Cranes stands at more than five hundred birds, not including those in the captive breeding population. These birds are all descendants of the original group of twenty-one back in 1941. This success is the result of international cooperation and is a testament to the hard work and incredible dedication of the conservationists who have fought to bring this species back from the brink of extinction.

Identification Whooping Cranes are white with red caps and dark facial markings. They have black wing tips and black beaks and feet. Juvenile birds have brown feathers mixed with white.

Vocalizations Whooping Cranes give a variety of calls, including flight intention, stress, unison, and guard calls. The loud trumpeting calls are generated by a five-foot-long trachea that coils within the sternum, giving the vocalizations much of their resonance and volume.

Nesting The nesting grounds are located in Wood Buffalo National Park, which straddles the border between Alberta and the Northwest Territories in Canada. Two eggs are typically laid, but usually only one young bird survives to fledge. For many years, biologists would remove the second egg and raise the hatchling in captivity to build the captive flock and increase the overall productivity of the species.

American Oystercatcher
(*Haematopus palliatus*)

CONSERVATION CONCERN SCORE: 15 (High)

OTHER DESIGNATIONS: 2021 USFWS Birds of Conservation Concern, 2021 Audubon
   Priority Birds list, US Shorebird Conservation Plan Species of High Conservation
   Concern (Category 4), State Protected (AL), State Special Concern (FL, GA, NC),
   Species of Greatest Conservation Need (MD, MS, SC, VA)

ESTIMATED POPULATION TREND 1966–2019: Not available

SIZE: Length 16–17 inches; wingspan 32 inches

American Oystercatchers often rest on jetties or docks when
they are not actively feeding. These birds were part of a larger
group that gathered near a boat launch in North Carolina.

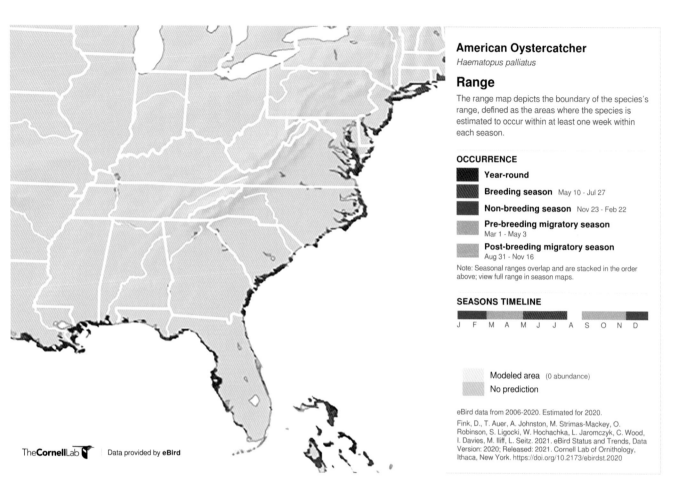

American Oystercatcher
Haematopus palliatus

Range

The range map depicts the boundary of the species's range, defined as the areas where the species is estimated to occur within at least one week within each season.

OCCURRENCE

Year-round

Breeding season   May 10 - Jul 27

Non-breeding season   Nov 23 - Feb 22

Pre-breeding migratory season
Mar 1 - May 3

Post-breeding migratory season
Aug 31 - Nov 16

Note: Seasonal ranges overlap and are stacked in the order above; view full range in season maps.

SEASONS TIMELINE

J F M A M J J A S O N D

Modeled area   (0 abundance)
No prediction

eBird data from 2006-2020. Estimated for 2020.
Fink, D., T. Auer, A. Johnston, M. Strimas-Mackey, O. Robinson, S. Ligocki, W. Hochachka, L. Jaromczyk, C. Wood, I. Davies, M. Iliff, L. Seitz. 2021. eBird Status and Trends, Data Version: 2020; Released: 2021. Cornell Lab of Ornithology, Ithaca, New York. https://doi.org/10.2173/ebirdst.2020

TheCornellLab | Data provided by eBird

Species Account American Oystercatchers are large, colorful, and vocal, making them hard to miss along the shores of the Atlantic Ocean and Gulf of Mexico. With their brilliant orange beaks and eye rings, yellow irises, black backs, and white bellies, oystercatchers are much more eye-catching than many of the other shorebirds of the region, which are typically shades of brown and gray that allow them to disappear into the surrounding sandscape. Strictly a coastal species, the American Oystercatcher is completely dependent on beaches, islands, intertidal flats, and salt marshes for its existence.

Oystercatchers actually do feed on oysters, although they take a variety of prey, including starfish, sea urchins, jellyfish, and marine worms. However, their preferred prey appears to be mollusks—soft-bodied invertebrates such as soft-shell and razor clams, blue mussels, ribbed mussels, and oysters. Because of their hard shells, these mollusks can be difficult for birds to eat, and oystercatchers are one of the few capable of doing so. Some species of gulls have learned how to crack the shells by flying their catch to a height and dropping it, but oystercatchers use a different method. Oystercatchers usually feed on tidal flats during low tide, visually searching for mollusks on the surface, in small pools, or on rocks. However, they are also capable of probing for prey beneath the surface of the wet sand with their long bills. Upon capture of a mollusk, the bird uses its laterally compressed, chisel-like bill to either cut the adductor chain that holds the shell closed or smash the shell with repeated hammering. Gulls, Willets, Ruddy Turnstones, and other birds frequently steal the oystercatchers' prey once they have removed the shell.

Scientists who conducted a comprehensive survey of the Atlantic and Gulf coasts in the winter of 2002–2003 estimated a population of eleven thousand American Oystercatchers in this region. Although the total population would have been higher, since this species also occurs along the Pacific coast, in Mexico, and in portions of coastal Central America, the Caribbean, and South America, such a small regional population was concerning. This led to extensive studies of the species. Recapture and observation of banded or marked birds revealed that the migration patterns of American Oystercatchers are complex. Most birds from the northern Atlantic coast move south for the nonbreeding season, often flying all the way to Florida. Birds that nest in the Carolinas or farther south may stay in the same area year-round or move north or south along the coast after the breeding season. Even birds from the same clutch have been shown to migrate in opposite directions. Mated pairs may select different migration strategies and spend the nonbreeding season in different areas, often hundreds of miles apart. In one instance, a pair that nested in North Carolina was separated by about five hundred miles, with one bird staying on its breeding territory and the other flying south to Florida.

Population trends for this species are difficult to estimate. However, with such a small overall population, the American Oystercatcher faces many threats moving forward. These include loss of habitat due to coastal development, heavy human recreational use of the birds' preferred nesting habitat, an increase in predators that take eggs and chicks, and a rising sea level and more frequent storm events due to climate change. The American Oystercatcher Working Group has developed management recommendations, with the goal of increasing the Atlantic and Gulf coast population to one and a half times current levels. To accomplish this, the group recommends identifying and protecting important habitat, especially in South Carolina, where a large percentage of the population spends the nonbreeding season. Other key conservation actions include controlling predators near breeding areas, avoiding disturbances near nesting locations during breeding season, and creating additional habitat by using dredge spoil.

Identification American Oystercatchers are large, brightly colored shorebirds. Adults have a yellow iris, orange eye ring, long orange bill, and long, thick pinkish legs. The head, neck, and back are black, and the belly is white. In flight they show a thick white stripe down the wing and a white patch at the base of the tail.

Vocalizations Oystercatchers give a sharp, high-pitched *kleep* call, as well as a piping call from the ground or in the air as part of a courtship display that may involve nearby pairs in similar displays.

Nesting In Georgia, pairs begin defending their territory as early as January; farther north in Virginia, it may be late February or early March before breeding activities commence. Nests consist of simple scrapes in the substrate—usually sand or shells, although sometimes the birds nest in the high-tide wrack line in salt marshes. Two or three grayish eggs are laid, heavily speckled with gray or chocolate brown.

Flying through a heavy coastal fog, these American Oystercatchers exhibit the broad white stripe down the wing that is characteristic of the species.

Piping Plover
(*Charadrius melodus*)

CONSERVATION CONCERN SCORE: 15 (High)

OTHER DESIGNATIONS: American Bird Conservancy watch list (Red), 2021 Audubon Priority Birds list, US Shorebird Conservation Plan Highly Imperiled (Category 5), State Protected (AL), State Special Concern (GA), State Threatened (KY, TN), State Endangered (MD, MS), Species of Greatest Conservation Need (FL, NC, PA, SC, VA), Great Lakes population Federally Endangered, Atlantic population Federally Threatened

ESTIMATED POPULATION TREND 1966–2019: Not available

SIZE: Length 7 inches; wingspan 19 inches

A male Piping Plover approaches a female on the Atlantic coast breeding grounds. His chest is puffed out, and he walks toward her with stiff legs that he quickly flicks up from the ground to a ninety-degree angle in front of him. As he gets closer, his feet will touch her back during this stiff walking display, just prior to copulation.

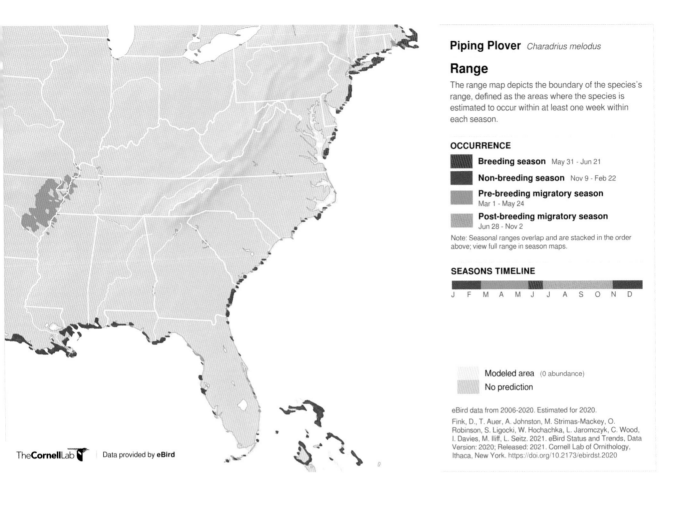

Piping Plover *Charadrius melodus*

Range

The range map depicts the boundary of the species's range, defined as the areas where the species is estimated to occur within at least one week within each season.

OCCURRENCE

Breeding season May 31 - Jun 21

Non-breeding season Nov 9 - Feb 22

Pre-breeding migratory season
Mar 1 - May 24

Post-breeding migratory season
Jun 28 - Nov 2

Note: Seasonal ranges overlap and are stacked in the order above; view full range in season maps.

SEASONS TIMELINE

J F M A M J J A S O N D

Modeled area (0 abundance)
No prediction

eBird data from 2006-2020. Estimated for 2020.
Fink, D., T. Auer, A. Johnston, M. Strimas-Mackey, O. Robinson, S. Ligocki, W. Hochachka, L. Jaromczyk, C. Wood, I. Davies, M. Iliff, L. Seitz. 2021. eBird Status and Trends, Data Version: 2020; Released: 2021. Cornell Lab of Ornithology, Ithaca, New York. https://doi.org/10.2173/ebirdst.2020

TheCornellLab | Data provided by **eBird**

Species Account With backs the color of sand, Piping Plovers blend into their surroundings and seem to disappear when they are not moving. Fortunately, they often make short runs along the beach before stopping again to search for food. These ghosts of the sand have an uncanny ability to blink in and out of sight, forcing observers to constantly hunt for birds they just saw a moment ago.

Their ability to hide in plain sight is not the only mysterious quality attributed to these pale, stocky shorebirds. For many years, scientists were unable to account for the full population of Piping Plovers during the nonbreeding season, despite extensive efforts to survey their known wintering grounds along the southern US Atlantic coast and the shores of the Gulf extending south into Mexico. It wasn't until 2012, when biologists discovered additional groups of birds wintering in the Bahamas, that this mystery was solved. This led to the creation of a 92,000-acre national park in the Bahamas to help protect the Piping Plover and other shorebirds.

The Piping Plover is somewhat unusual, in that there are three distinct breeding populations: one in the northern Great Plains extending into Canada, a federally endangered population that breeds in the Great Lakes region, and a federally threatened population that breeds along the shores of the Atlantic from North Carolina or Virginia north into the Canadian Maritime Provinces. Some have argued that there are observable physical differences among the three populations, such as the extent of the black band across the chest, that could be used to separate these populations into different species. Although others dispute the extent of the physically distinguishing characteristics, the three populations apparently have different migration patterns, with Great Plains populations wintering on the Gulf coast, Great Lakes birds wintering in South Carolina and Georgia, and Atlantic coast birds spending the winter in North Carolina and the Caribbean. However, there is still much to learn about the migration patterns of these birds, and there may be geographic overlap in some cases. Genetic testing indicates some differences among the three populations as well, with mitochondrial DNA evidence showing that the Great Lakes birds are more genetically similar to the Great Plains population. Currently, two subspecies are recognized: *C. melodus melodus,* which includes the birds of the Atlantic coast, and *C. melodus circumcinctus,* which includes the Great Lakes and Great Plains birds.

The Piping Plover's diet consists largely of insects, crustaceans, marine worms, and other invertebrates. The inland Great Plains birds are more dependent on insects, including beetles, midges, flies, and other species. Piping Plovers hunt by running a short distance and then stopping and pecking at the sand to capture their quarry. Their short, stubby beaks are surprisingly powerful, capable of crushing prey or inflicting serious injury on other birds that enter their territories.

The Piping Plover has likely experienced severe declines over the past hundred years or more. Causes include loss of coastal beach habitat, an increase in predators, and significant human disturbance during nesting and nonbreeding seasons. Efforts to control predators and rope off portions of popular beaches during the nesting season have helped protect this species, resulting in higher reproduction success. Some populations have apparently increased in recent years, in part because of these protections. Based on the most recent survey, the total global Piping Plover population across all three breeding groups is estimated at eight thousand birds.

Identification The Piping Plover is a small, stocky shorebird that often holds its body parallel to the ground. The back is brownish gray, and the belly is white. During the breeding season, the male has a black bar across the forehead, a single black ring around the neck, an orange beak with a black tip, and orange legs. The female is similar, but usually with less bold black markings. During the nonbreeding season, the beak becomes black, the black markings disappear from the forehead altogether, and the black ring fades to an incomplete band of light gray.

Vocalizations Piping Plovers give a hollow, whistled *peep-lo* call, with the second note at a lower pitch than the first. Males also give a piping call during courtship flights over their territory. A variety of other contact and alarm calls are also heard.

Nesting Male Piping Plovers have elaborate breeding behaviors that include parallel running displays with neighboring males at the edges of their territories. When a male approaches his mate, he performs a high-stepping display with a puffed chest, his feet rapidly kicking out parallel to the ground in front of him. Nests are simple scrapes in the sand, sometimes lined with small pebbles. Nests are often located near Least Tern colonies and typically contain a clutch of four buff-colored eggs with darker brown speckling.

An adult Piping Plover on the Gulf coast in nonbreeding plumage, showing a total lack of black on the forehead and no neck ring.

An adult Piping Plover nearing full breeding plumage
scans a Gulf coast mudflat in search of prey to fuel
its migration north to the breeding grounds.

Wilson's Plover
(*Charadrius wilsonia*)

CONSERVATION CONCERN SCORE: 16 (High)

OTHER DESIGNATIONS: American Bird Conservancy watch list (Yellow),
2021 USFWS Birds of Conservation Concern, US Shorebird Conservation Plan
Species of High Conservation Concern (Category 4), State Protected (AL),
State Special Concern (GA, NC), State Endangered (MD, VA),
Species of Greatest Conservation Need (FL, MS, SC)

ESTIMATED POPULATION TREND 1966–2019: Not available

SIZE: Length 8 inches; wingspan 19 inches

Crabs make up a good portion of the diet of the Wilson's Plover. Its
heavy beak is helpful in dismembering crabs before consuming them.

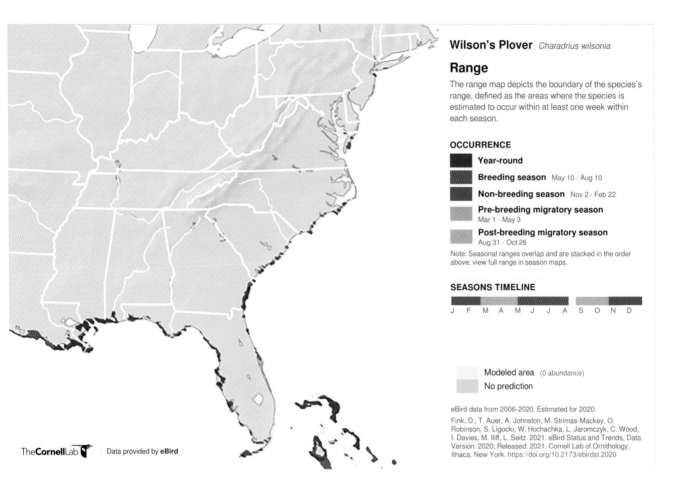

Wilson's Plover *Charadrius wilsonia*

Range

The range map depicts the boundary of the species's range, defined as the areas where the species is estimated to occur within at least one week within each season.

OCCURRENCE

- Year-round
- Breeding season   May 10 - Aug 10
- Non-breeding season   Nov 2 - Feb 22
- Pre-breeding migratory season   Mar 1 - May 3
- Post-breeding migratory season   Aug 31 - Oct 26

Note: Seasonal ranges overlap and are stacked in the order above; view full range in season maps.

SEASONS TIMELINE

J F M A M J J A S O N D

Modeled area   (0 abundance)
No prediction

eBird data from 2006-2020. Estimated for 2020.

Fink, D., T. Auer, A. Johnston, M. Strimas-Mackey, O. Robinson, S. Ligocki, W. Hochachka, L. Jaromczyk, C. Wood, I. Davies, M. Iliff, L. Seitz. 2021. eBird Status and Trends, Data Version: 2020; Released: 2021. Cornell Lab of Ornithology, Ithaca, New York. https://doi.org/10.2173/ebirdst.2020

TheCornellLab | Data provided by **eBird**

Species Account   This is how Arthur Cleveland Bent described the Wilson's Plover in 1929:

> Here, if we love to wander in these seaside solitudes, we may see this gentle bird running along the beach ahead of us, his feet twinkling so fast we can hardly see them; he is unafraid, as he stops and turns to watch us; the black bands on his head and chest help to obliterate his form and he might be mistaken for an old sea-shell or a bit of driftwood; but, as we draw near, he turns and runs on ahead of us, leading us thus on and on up the beach. There is an air of gentleness in his manner and an air of wildness in his note as he flies away.

Originally described for science by George Ord, the Wilson's Plover is named for Ord's friend Alexander Wilson, who collected the first specimen on the southern coast of New Jersey in May 1813. Today, the Wilson's Plover rarely occurs that far north; it is more typically found from Virginia south through Florida and the Gulf coast. Four subspecies exist, with populations along coastlines throughout the Caribbean, the

Pacific and Atlantic coasts of Mexico and Central America, and coastal areas of South America as far south as Brazil and Peru.

Although Wilson's Plovers take small numbers of insects, shrimp, and worms, fiddler crabs are their main food source, accounting for up to 98 percent of their diet in some cases. Wilson's Plovers forage heavily day or night during falling tides, when the crabs are more exposed. They are visual hunters, stalking along the beach before running and extending their necks to capture a crab. The birds shake the crabs and often remove the legs before consuming the body. The beak, which is heavy for a shorebird, may be especially helpful in taking the crustaceans apart.

Courtship behavior during the breeding season begins with the male creating several scrapes in his territory. If a female approaches, the male steps out of the scrape, lowers and fans his tail while raising one wing and lowering the other, and points his bill toward the scrape. Nests are often placed in sand, crushed shells, or gravel, with a clump of vegetation or some small topographic feature providing a windbreak. Breeding pairs are monogamous during the nesting season. An individual pair typically defends a territory between three hundred and three thousand feet in diameter. Although the Wilson's Plover is not a colonial nesting species, the birds often place their nests within sight of other plover nests. When a predator approaches, multiple pairs may cooperate to defend the general area.

Although the Breeding Bird Survey does not have reliable data on population trends for this species, other studies have shown that its range in the United States is contracting and overall population numbers are declining. Audubon Christmas Count data reflect a 78 percent decline in populations wintering in the United States over the last forty years. The US Shorebird Conservation Plan lists Wilson's Plover as a species of high concern, based on known threats to its breeding and nonbreeding grounds, low overall population size, and limited geographic range. Current estimates put the total global population at fewer than thirty-two thousand birds. In the United States, fewer than eighty-six hundred individuals remain, with the largest numbers occurring in Texas and Louisiana. Virginia, Mississippi, and Alabama have only a couple dozen nesting pairs each, but as many as 670 pairs may still nest in the Carolinas.

Threats include human disturbance of nesting locations above the high-tide line on heavily trafficked beaches. Off-road vehicles have been known to run over this species. Foot traffic, vehicles, and pets can also destroy nests, contribute to low productivity, or cause nest abandonment. Further development of coastal habitats and climate change are likely the greatest threats to the remaining population.

Identification Wilson's Plovers have plain brown backs with white underparts. The bill is all black and thicker than that of most other shorebirds of similar size. The legs are pinkish gray. In the breeding season, the male has a thick black band across the upper breast, a black line extending from the eye to the beak, and a black mark on the forehead. Females and birds in nonbreeding plumage have more brown than black in the breast band and on the head.

Vocalizations   Simple calls include a variety of short, rattling notes, as well as some slurred whistles.

Nesting   Typically, three oval eggs are laid. The eggs are buff colored with darker brown markings. Incubation lasts approximately twenty-five days. In hot weather, the parents may stand and shade the eggs or soak their belly feathers to cool the eggs when sitting on the nest. Chicks can walk within one to two hours after hatching and leave the nest to hide in nearby vegetation shortly thereafter.

An adult male Wilson's Plover in breeding plumage.

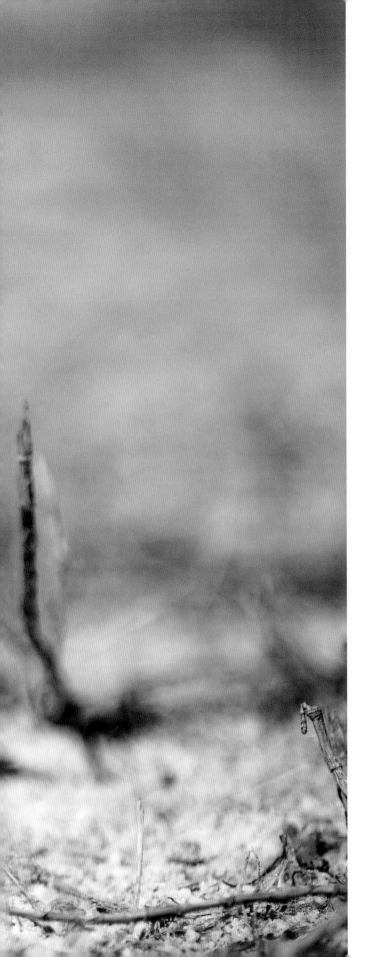

An adult Wilson's Plover showing the typical brown neck band of its nonbreeding plumage. Its heavy bill compared to that of other shorebirds helps identify this species, even at a distance.

Snowy Plover
(*Charadrius nivosus*)

CONSERVATION CONCERN SCORE: 15 (High)

OTHER DESIGNATIONS: American Bird Conservancy watch list (Yellow), 2021 USFWS Birds of Conservation Concern, US Shorebird Conservation Plan Highly Imperiled (Category 5), State Protected (AL), State Threatened (FL), State Endangered (MS), Pacific coast population Federally Threatened (as of 1993)

ESTIMATED POPULATION TREND 1966–2019: Not available

SIZE: Length 6 inches; wingspan 17 inches

The sand-colored plumage of the Snowy Plover makes the birds almost invisible on beaches and barrier islands. This adult bird is transitioning from nonbreeding to full breeding plumage.

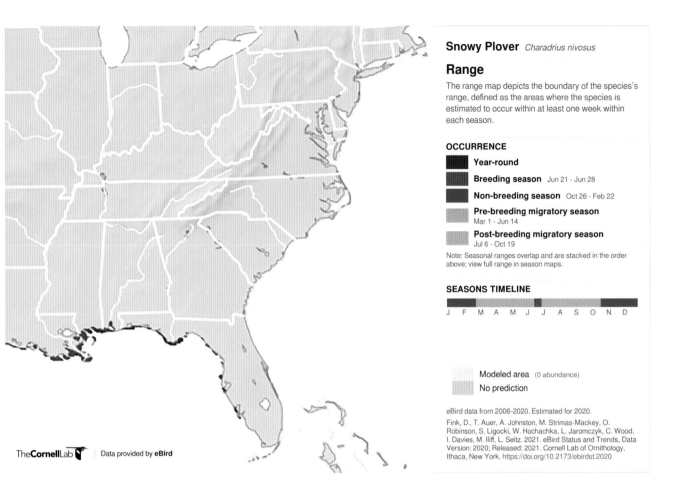

Snowy Plover *Charadrius nivosus*

Range

The range map depicts the boundary of the species's range, defined as the areas where the species is estimated to occur within at least one week within each season.

OCCURRENCE

■ **Year-round**

■ **Breeding season** Jun 21 - Jun 28

■ **Non-breeding season** Oct 26 - Feb 22

■ **Pre-breeding migratory season** Mar 1 - Jun 14

■ **Post-breeding migratory season** Jul 6 - Oct 19

Note: Seasonal ranges overlap and are stacked in the order above; view full range in season maps.

SEASONS TIMELINE

J F M A M J J A S O N D

Modeled area (0 abundance)

No prediction

eBird data from 2006-2020. Estimated for 2020.

Fink, D., T. Auer, A. Johnston, M. Strimas-Mackey, O. Robinson, S. Ligocki, W. Hochachka, L. Jaromczyk, C. Wood, I. Davies, M. Iliff, L. Seitz. 2021. eBird Status and Trends, Data Version: 2020; Released: 2021. Cornell Lab of Ornithology, Ithaca, New York. https://doi.org/10.2173/ebirdst.2020

TheCornellLab | Data provided by **eBird**

Species Account Snowy Plovers are widely distributed, with breeding populations stretching from Saskatchewan in Canada, throughout much of the arid West, to central Mexico, the Caribbean, and parts of the Pacific coast of South America. Despite this broad range, Snowy Plovers have disappeared from many places where they once nested. In the Appalachians and the southeastern United States, Snowy Plovers can be found year-round along portions of the Gulf coast in Mississippi, Alabama, and Florida. The population in Florida appears to be at least partially migratory, as there are about 50 percent more birds during the breeding season than in the winter.

When feeding, Snowy Plovers on the Gulf coast run and then pause as they search for prey. They occasionally shake a foot rapidly over the surface of water, mud, or sand to reveal hidden prey. Their diet includes small crustaceans, marine worms, and insects. Snowy Plovers have been observed charging into a dense gathering of insects sitting on the surface of the sand. As the insects take flight, the plover snaps at them with its beak to capture and consume them. At inland breeding locations across the West, insects make up a greater portion of the diet, including larvae the birds probe for in the ground.

Snowy Plovers in Florida begin pair formation as early as January and usually begin laying eggs in March. The newly fledged young are unusually mobile. They leave the nest site a few hours after hatching and are capable of capturing their own food immediately. The parents do not feed the young but simply lead them to appropriate feeding areas. The adult birds defend the chicks from predators and brood them to provide warmth. In western populations, the adult female often leaves the chicks in the care of the male while she moves to a new area to begin a second nest, sometimes hundreds of miles away. On the Gulf coast, both parents usually stay with the chicks until they are able to fly.

A number of factors have contributed to the overall decline in the population of this species, especially loss of habitat due to coastal development and coastal recreation leading to decreased nesting success. Entanglement with discarded fishing line can cause the loss of toes or feet or even death in some cases. Pollution, including oil spills, is an ongoing threat. Some estimates put the total population of Snowy Plovers in the United States and the Pacific coast of the Baja peninsula in Mexico at eighteen thousand birds. The population on the Gulf coast is much smaller. Surveys in the late 1990s and early 2000s found that wintering birds numbered only 13 in Mississippi, none in Alabama, and 311 in Florida. The number of birds in the western Gulf coast was higher, with 690 birds in Texas and 1,191 in the Laguna Madre region of Tamaulipas, Mexico. Given such small populations, it is important to minimize threats to the species. Recreational activities by human beachgoers can cause adult birds to leave the nest or waste energy running or flying away. Off-road vehicles on beaches can damage nesting habitat and kill both Snowy Plover adults and chicks. Unleashed dogs near nesting areas can also kill adult birds and chicks. Temporarily fencing off nesting areas during the breeding season and posting educational signs have been effective in improving this species' breeding success. Volunteer patrols have also been successful at reducing the amount of disturbance at known nesting sites. Predator exclusion cages have been used experimentally in an attempt to increase hatching success, and although hatching rates improved, overall fledging success declined, suggesting that these cages may be increasing adult mortality or lowering overall chick survival in other ways.

Identification Snowy Plovers have pale gray backs and white bellies. Breeding birds sport black markings on the forehead and behind the eye and an incomplete black neck ring. In the nonbreeding season, these black markings fade entirely or become brownish gray. The legs and beak are grayish black in all seasons, helping to differentiate the Snowy Plover from the similar Piping Plover, which has orange on the beak and feet.

Vocalizations Both sexes give a *prrrt* call. Males also give a *towheet* call to advertise for a mate and a *churr* call when defending their territory against other males.

Nesting The nest is a simple scrape in the sand, sometimes in close proximity to the nests of other Snowy Plover pairs. The nest scrape is often near a piece of driftwood or clump of grass, which may protect the nest or nestlings from drifting sand during windy weather. Three pale, sand-colored, heavily speckled eggs are laid. Larger clutches have been reported, which may be the result of two females laying their eggs in the same scrape.

Red Knot
(*Calidris canutus*)

CONSERVATION CONCERN SCORE: 12 (Moderate)

OTHER DESIGNATIONS: American Bird Conservancy watch list (Yellow),
2021 Audubon Priority Birds list, US Shorebird Conservation Plan Species of High
Conservation Concern (Category 4), State Protected (AL), State Special Concern
(GA), State Threatened (MD, VA), Species of Greatest Conservation Need
(FL, MS, NC, PA, SC, TN), *C. c. rufa* subspecies Federally Threatened

ESTIMATED POPULATION TREND 1966–2019: Not available

SIZE: Length 11 inches; wingspan 23 inches

Red Knots in nonbreeding plumage are rather plainly
colored, but their stocky profile helps identify them.

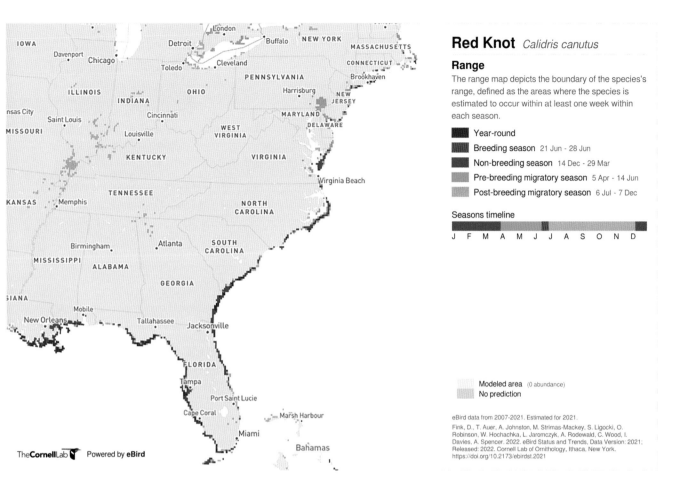

Red Knot *Calidris canutus*

Range

The range map depicts the boundary of the species's range, defined as the areas where the species is estimated to occur within at least one week within each season.

Year-round
Breeding season 21 Jun - 28 Jun
Non-breeding season 14 Dec - 29 Mar
Pre-breeding migratory season 5 Apr - 14 Jun
Post-breeding migratory season 6 Jul - 7 Dec

Seasons timeline

J F M A M J J A S O N D

Modeled area (0 abundance)
No prediction

eBird data from 2007-2021. Estimated for 2021.
Fink, D., T. Auer, A. Johnston, M. Strimas-Mackey, S. Ligocki, O. Robinson, W. Hochachka, L. Jaromczyk, A. Rodewald, C. Wood, I. Davies, A. Spencer. 2022. eBird Status and Trends, Data Version: 2021; Released: 2022. Cornell Lab of Ornithology, Ithaca, New York. https://doi.org/10.2173/ebirdst.2021

TheCornellLab Powered by eBird

Species Account Perfectly built for long-distance migration, Red Knots are one of the true marvels of the natural world. Weighing less than a cup of coffee, this remarkable bird spends its winters as far south as Tierra del Fuego at the southern tip of South America. As breeding season approaches, the birds migrate north to the Arctic Circle to nest—only to fly south again after the short breeding season. This amazing journey means that some birds travel eighteen thousand miles in a single year, much of that distance over the open Atlantic Ocean. To fuel their flight, knots have to pack in an incredible number of calories during several short stopovers along the way. The birds return to the same stopover sites year after year, and in many cases they arrive emaciated and in need of immediate sustenance after a nonstop flight of fifteen hundred miles or more. At these stopover sites, birds can increase their body weight by as much as 10 percent per day when conditions are good. They sometimes stay long enough to double their body weight before attempting the next segment of the journey.

Perhaps the most critical stopover site for this species is along Delaware Bay on the East Coast of the United States. This site is home to the largest concentration of nesting horseshoe crabs in the world, and the crabs' eggs are the perfect food for migrating Red Knots. Prior to 1980, the number of Red Knots using Delaware Bay during the month of May averaged between 100,000 and 150,000 birds. Large-scale commercial harvesting of horseshoe crabs in Delaware Bay began in the 1990s, and today the average number of knots using the area in May is fewer than 26,000 birds. Up to 90 percent of the world population of the *rufa* subspecies of Red Knot can be found on the beaches of Delaware Bay on a single day in May. Estimates suggest a 75 percent drop in numbers of the *rufa* subspecies between 2000 and 2016. Although the population stabilized somewhat beginning in 2010, the most recent surveys again show significant declines. In fact, a 2021 survey found only 6,880 birds on the shores of Delaware Bay. Horseshoe crab harvests are now carefully managed to help stabilize and restore Red Knot populations, but given the recent drop in knot numbers, there are renewed calls to end crab harvesting entirely. In all, there are five subspecies of the Red Knot worldwide, three of which occur in the United States.

Red Knots are somewhat unusual among shorebirds in that, for much of the year, they eat mussels and mollusks—shells and all. In preparation for migration, however, the knot's flight muscles increase in size but their stomachs and gizzards shrink. Because the gizzards normally help digest food, the birds are forced to search for softer prey during migration, which is why horseshoe crab eggs are so critical for the birds' survival. Upon reaching the frozen Arctic nesting grounds in June, the birds must eat seeds for a short time until insects and invertebrates become available; otherwise, they have to survive on fat reserves from their last stopover. This makes stopover habitat like Delaware Bay even more important because it may determine how long the birds can survive in the Arctic until food becomes abundant there.

Because of recent declines in the *rufa* subspecies, in July 2021 the US Fish and Wildlife Service proposed designating nearly 650,000 acres as "critical habitat" for the Red Knot across a number of states where it is found during migration or occasionally during the winter. In the Southeast, this includes Alabama, Mississippi, Florida, Georgia, South Carolina, North Carolina, and Virginia. This proposed designation is designed to protect existing habitat and focus conservation efforts where it is most needed to help this population survive.

Identification Red Knots are the largest of the peeps (a group of small shorebirds) in North America and are about the size of a robin. They are long-winged and look streamlined in flight. In breeding plumage (May–August), they have a salmon-colored breast, throat, and eye line with a gray-, black-, and salmon-dappled back. The black bill is relatively short and straight. In nonbreeding plumage (September–April), adults have a gray back and chest with barring down the flanks and a white breast. Juveniles look similar to nonbreeding adults, but the feathers of the back are edged in white, giving them a scaly appearance. Birds in nonbreeding plumage can be mistaken for the smaller Dunlin, whose beak is longer and droops at the tip.

Vocalizations Knots are usually silent, although they may give a low *wet-wet* call. Males on the breeding grounds give a soft, wailing *quer-wer* call as they fly in high circles over their territory.

Nesting The nest is a cup-shaped depression on the ground in the open tundra, usually near water. Eggs are faint olive with darker markings concentrated toward the larger end. Young leave the nest shortly after hatching and are tended by both parents for a short time. The female leaves to begin her migration southward before the young are able to fly. The first nest of this species was not recorded until 1909, during Robert Peary's expedition to the North Pole.

This adult Red Knot in nearly full breeding plumage pauses to bathe after feeding on horseshoe crab eggs and refueling for its journey north.

While transitioning from nonbreeding to breeding plumage, Red Knots develop a
patchy appearance, as salmon-colored feathers replace the simpler grays and whites.

In flight, Red Knots are streamlined and appear long winged. Perfectly designed for long-distance migration, they are true marvels of the natural world.

Short-billed Dowitcher
(*Limnodromus griseus*)

CONSERVATION CONCERN SCORE: 14 (High)

OTHER DESIGNATIONS: 2021 USFWS Birds of Conservation Concern, US Shorebird Conservation Plan Species of High Conservation Concern (Category 4), Species of Greatest Conservation Need (FL, KY, MD, SC, VA)

ESTIMATED POPULATION TREND 1966–2019: Not available

SIZE: Length 11 inches; wingspan 19 inches

In nonbreeding plumage, Short-billed Dowitchers are gray overall, with long bills, long yellowish-green legs, and dark speckles on the white undertail coverts.

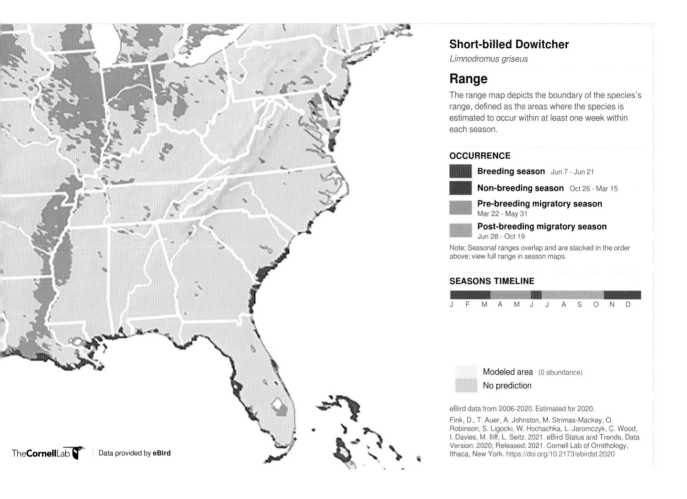

Short-billed Dowitcher
Limnodromus griseus

Range

The range map depicts the boundary of the species's range, defined as the areas where the species is estimated to occur within at least one week within each season.

OCCURRENCE

	Breeding season Jun 7 - Jun 21
	Non-breeding season Oct 26 - Mar 15
	Pre-breeding migratory season Mar 22 - May 31
	Post-breeding migratory season Jun 28 - Oct 19

Note: Seasonal ranges overlap and are stacked in the order above; view full range in season maps.

SEASONS TIMELINE

J F M A M J J A S O N D

	Modeled area (0 abundance)
	No prediction

eBird data from 2006-2020. Estimated for 2020.
Fink, D., T. Auer, A. Johnston, M. Strimas-Mackey, O. Robinson, S. Ligocki, W. Hochachka, L. Jaromczyk, C. Wood, I. Davies, M. Iliff, L. Seitz. 2021. eBird Status and Trends, Data Version: 2020; Released: 2021. Cornell Lab of Ornithology, Ithaca, New York. https://doi.org/10.2173/ebirdst.2020

TheCornellLab | Data provided by **eBird**

Species Account In the mid-1800s the Short-billed Dowitcher was incredibly numerous. In May 1868 a flock of birds that was likely this species was described on the coast of Maine: It "extended from horizon to horizon, and . . . lasted over three hours. . . . The body of birds must have been twelve or fifteen miles wide and at least one-hundred long . . . [and they flew] in bunches from a dozen to several hundred and were visible in all directions" (Palmer 1949). Also in the mid-1800s, Alexander Wilson reported "eighty-five . . . taken at the discharge of one musket" (Forbush 1912). Because of extensive shooting for sport and for food, these massive flocks were gone by the late 1800s. By the early 1900s, some authors speculated that the species might be nearing extinction. With the implementation of migratory bird protection, the species partially recovered, but populations are now believed to be on the decline again.

The name *dowitcher* is thought to come from an early reference to the German Snipe, or *Duitschers,* in the Pennsylvania Dutch dialect. Also known as the Red-breasted Snipe by early ornithologists, the Short-billed Dowitcher is slightly smaller and less colorful

This Long-billed Dowitcher is transitioning from nonbreeding plumage to its more colorful breeding plumage and is shown for comparison to the very similar Short-billed Dowitcher.

than the Long-billed Dowitcher, which prefers inland wetlands to the saltwater marshes and coastal beaches visited by the Short-billed species during migration. Despite its name, the Short-billed Dowitcher has a long bill and uses sensors in its beak to probe mudflats and other wet soils for prey. It is also capable of feeding on the surface, searching day or night for aquatic worms, insects, mollusks, shrimp, fiddler crabs, horseshoe crab eggs, and seeds. In a diet study from the Atlantic coast, Short-billed Dowitchers fed on 88 percent animal matter and 12 percent plant material (Sperry 1940). On the breeding grounds, birds consume insect larvae, freshwater snails, spiders, beetles, and mosquitoes.

The Short-billed Dowitcher was overlooked and misunderstood for many years. In fact, until the 1930s, scientists thought Short-billed and Long-billed Dowitchers were the same species. There are now three recognized subspecies of Short-billed Dowitcher, and each has its own separate breeding area and migration pattern. *L. g. griseus* nests in eastern Canada, migrates along the Atlantic coast, and winters along the coastline of the United States and as far south as Brazil. *L. g. hendersoni* breeds along the shores of Hudson Bay and west through northern Manitoba, Alberta, and Saskatchewan; it winters in the southeastern United States and south to Panama. *L. g. caurinus* breeds in Alaska, the Yukon, and northwest British Columbia and winters along the Pacific Ocean from California south to Peru.

In flight, dowitchers are recognized by their long bills and the black and white patterns covering their tails and the inverted V–shaped portion of their rumps.

Although Breeding Bird Survey data do not exist for this species, other surveys conducted along the Atlantic coast during migration indicate a decline of 46 percent between 1972 and 1983. Similar surveys conducted along the Canadian coastline show significant declines from 1974 to 1991. The most recent estimates put the global population of this species (including all three subspecies) between 150,000 and 320,000 birds.

Identification In breeding plumage, Short-billed Dowitchers typically show more white on the belly and more of an orange color compared with the darker chestnut tones of Long-billed Dowitchers. In nonbreeding season, Short-billed Dowitchers are a cool gray overall, with white bellies, lightly speckled flanks, a faint white eyebrow stripe, greenish legs, and a long, bicolored bill that is darker toward the tip.

Vocalizations Short-billed Dowitchers give a soft *tu-tu-tu* call. This is the best way to distinguish this species from the very similar Long-billed Dowitcher, which gives a short, reedy *keek*.

Nesting Nest sites are constructed in bog or tundra habitats, often near water. Four eggs are typically laid, and they are olive with darker brown markings. Both sexes participate in incubation.

Willet
(*Tringa semipalmata*)

CONSERVATION CONCERN SCORE: 14 (High)

OTHER DESIGNATIONS: 2021 USFWS Birds of Conservation Concern, US Shorebird
   Conservation Plan Species of Moderate Conservation Concern (Category 3),
   State Protected (AL), Species of Greatest Conservation Need (FL, MD, NC, SC, VA)

ESTIMATED POPULATION TREND 1966–2019: −28%

SIZE: Length 15 inches; wingspan 26 inches

An adult Willet in full breeding plumage flashes its
large black and white wing patches as it lands and
prepares to feed in a shallow coastal wetland.

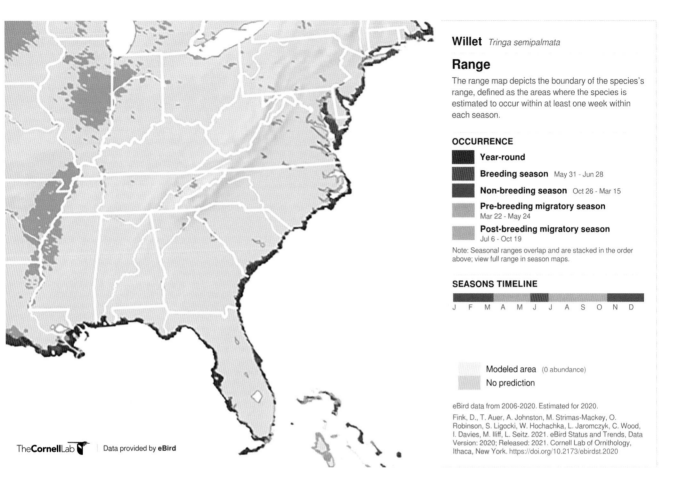

Species Account Although the Willet is found in many habitats, its shrill, rolling call evokes the sound of the pounding surf and the feel of the sun on the skin and sand between the toes. They may be seen at inland lakes, reservoirs, and wetlands during migration, but Willets are typically found most reliably and in greatest numbers along the shore, which, in the Southeast, means the Atlantic and Gulf coasts.

When standing, the Willet is a large, plain, rather nondescript bird. However, when it raises its wings in display or to take flight, striking black and white wing stripes are revealed. In addition to the birds in the Southeast, there is a western population that breeds in the prairie marshes of the northern Great Plains and in portions of the interior West. Some individuals from this western population winter along the Pacific coast from Oregon to Peru, while others apparently head south and east to winter with the resident birds along the Atlantic coast. Some birds travel to the Caribbean and as far south as Brazil. The western population of Willets is slightly larger and paler than the eastern population, and there is evidence that their vocalizations differ slightly as well.

Willets have long, straight beaks with sensitive tips that help them locate their prey. Because they hunt by feel as well as by sight, Willets can feed both night and day. Their diet consists largely of small fish, worms, crabs, clams, aquatic beetles, spiders, and other aquatic invertebrates found in the wet sand along the waterline.

Willets were extensively hunted during the 1800s and early 1900s, and this pressure undoubtedly contributed to the overall decline of the species. Willet eggs were prized as food, and Audubon's (1827) *Birds of America* reported that the young "grow rapidly, become fat and juicy, and by the time they are able to fly, afford excellent food." Hunting of the birds in the United States ceased with the passage of new laws in the early 1900s, and Willet populations began to rise. In the last several decades, however, populations have started to decline again, likely due to human disturbance and loss of habitat as coastlines are developed. Conversion of habitat to agriculture and oil and gas development are likely the two major contributors to the decline of Willets in the West.

Identification Willets are large, long-legged shorebirds with straight, relatively heavy, bicolored bills that shade toward black at the tip. The large, broad wings have bold black and white stripes that are evident in flight in all plumages. Eastern birds in breeding plumage are heavily barred on the head, breast, back, and flanks, with white underparts and grayish legs and bill. Western birds in breeding plumage are lighter overall, with only sparse markings on the throat and flanks. Both populations are uniformly light gray in the nonbreeding season. All Willets show a white patch at the base of the short tail in flight, with darker markings at the tip. Yellowlegs are similar to the Willet except for their bright yellow legs.

Vocalizations The call is a loud, rolling *pill will willet, pill will willet,* often given in flight. Also heard are a short, repeated *wik* alarm call and a harsh *kuk kuk kuk.*

Nesting The male performs aerial flight displays, calling loudly and fluttering his wings. He creates a shallow scrape two inches deep, but the female completes the nest, which is usually located among dense, short grass or sometimes on open ground. The nest is lined with grasses and is often well hidden—partially under vegetation. Both sexes incubate the eggs, but only the male incubates at night. Four or five olive-buff eggs with irregular brown splotches are laid. To distract would-be predators from the nest, adults may drag a wing.

These Willets are exhibiting pair-bonding behavior
during the early stages of courtship.

Willets call frequently in flight, and their *pill will willet* is a distinctive sound along the Atlantic and Gulf coasts.

Least Tern
(*Sternula antillarum*)

CONSERVATION CONCERN SCORE: 14 (High)

OTHER DESIGNATIONS: American Bird Conservancy watch list (Red),
   2021 USFWS Birds of Conservation Concern, 2021 Audubon Priority Birds list,
   State Protected (AL), State Special Concern (GA, NC), State Threatened (FL, MD),
   State Endangered (KY, MS), Species of Greatest Conservation Need (SC, TN, VA)

ESTIMATED POPULATION TREND 1966–2019: −77%

SIZE: Length 9 inches; wingspan 20 inches

Least Terns nest along sandy beaches with sparse vegetation.
Islands are ideal sites because of the lack of mammalian predators.

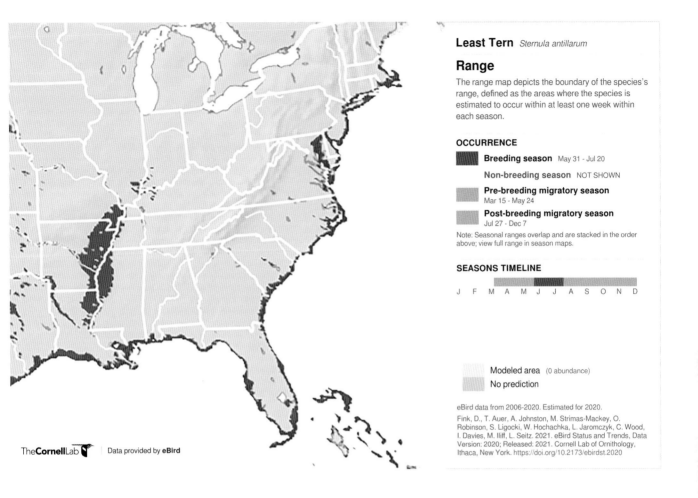

Least Tern *Sternula antillarum*

Range

The range map depicts the boundary of the species's range, defined as the areas where the species is estimated to occur within at least one week within each season.

OCCURRENCE

Breeding season May 31 - Jul 20

Non-breeding season NOT SHOWN

Pre-breeding migratory season
Mar 15 - May 24

Post-breeding migratory season
Jul 27 - Dec 7

Note: Seasonal ranges overlap and are stacked in the order above; view full range in season maps.

SEASONS TIMELINE

J F M A M J J A S O N D

Modeled area (0 abundance)
No prediction

eBird data from 2006-2020. Estimated for 2020.
Fink, D., T. Auer, A. Johnston, M. Strimas-Mackey, O. Robinson, S. Ligocki, W. Hochachka, L. Jaromczyk, C. Wood, I. Davies, M. Iliff, L. Seitz. 2021. eBird Status and Trends, Data Version: 2020; Released: 2021. Cornell Lab of Ornithology, Ithaca, New York. https://doi.org/10.2173/ebirdst.2020

TheCornellLab   Data provided by **eBird**

Species Account Least Terns are the smallest tern species in North America. They nest on sandy beaches, on dredged islands, or in other habitats with little vegetation along coastlines or major inland rivers. Least Tern populations have fluctuated significantly over time. In the late 1800s and early 1900s their feathers were popular in the hat-making trade, and their eggs were collected for food. The population was critically low when the Migratory Bird Treaty Act of 1918 was passed, which helped protect this and many other species. Although Least Tern numbers initially rebounded, more recent declines have occurred due to the creation of dams and artificial water levels and flows on large rivers, which can flood nests or destroy habitat; human disturbance on beaches during nesting season; and coastal development. Breeding Bird Survey data reflect a decline of roughly 77 percent since 1966, but few BBS routes cover Least Tern habitat. Other sources of information seem to confirm that serious declines have taken place in coastal populations, despite efforts to protect Least Tern colonies on public beaches during nesting season and to create additional habitat by building dredge spoil islands and artificial nesting areas on rooftops or in parking lots. The news is better for the Interior

Least Tern, a subspecies that nests on islands in major rivers across a swath of the central United States. In January 2021 the Interior Least Tern had recovered sufficiently to be removed from the federal Endangered Species List. Current estimates place the total number of individuals in this subspecies at eighteen thousand.

The Least Tern is a migratory species, breeding on the Atlantic, Gulf, and Pacific coasts and along major interior rivers such as the Mississippi. Birds first arrive along the Gulf coast in late March or early April. By mid-April, interior populations have made it to their nesting grounds as far north as Kentucky. Fall migration is more drawn out. It begins as early as June in some places, and many coastal colonies depart in August. Stragglers have been seen as late as October in Kentucky. Some Least Terns migrate to the Gulf in southern Mexico; others fly south over the open ocean through the Caribbean, where they winter in the islands there or continue farther south to the northern coast of South America.

Least Terns hunt by hovering low over shallow water and then diving in and snatching the prey in their beaks. It appears that the size of the fish is more important than the species. Least Terns reportedly consume more than fifty different species of small fish, the most common being anchovies, menhaden, killifish, shad, and shiners. The prey fish usually measure three-quarters of an inch to three and a half inches long, but the size of the fish taken during nesting season depends on the size of the chicks the parents are feeding. Other prey includes shrimp and a variety of insects captured on the wing or on the ground or scooped from the water's surface, adding variety to the diet.

Fierce defenders of their nesting colonies, Least Terns mob intruding gulls, crows, and other species perceived to be a threat. They are buoyant, graceful fliers and often pursue and dive at potential predators, sometimes striking persistent intruders with their bills or their feet to drive them away from the nest. These efforts are not as effective at deterring mammalian predators, which include coyotes, skunks, raccoons, feral hogs, foxes, and feral dogs and cats. Predation often results in low productivity in Least Tern colonies, and heavy predation or human disturbance can cause colony abandonment.

Identification   Least Terns are white below, with gray backs. Their bright yellow beaks and yellow legs distinguish them from all other North American tern species. In flight, their wings are narrow and swept back. During the breeding season, the birds have a black cap on the head and a bright white spot on the forehead, which is separated by a strong black line running from the beak to the black cap.

Vocalizations   Common calls include a sharp, grating *kip kip kip kideek* and shrill, two-syllable alarm notes when defending the colony.

Nesting   Nests are shallow scrapes on the ground in loose sand, shells, or gravel. The eggs are buffy, with dark brown speckles and splotches. A typical clutch consists of one to three eggs. In hot weather, the adult terns may fly to the water, wet their belly feathers, and then return to the nest to cool the eggs. Typically, a single brood is produced each year, but Least Terns along the Gulf coast may raise two broods.

This juvenile Least Tern was capable of flight but was still being tended by one of the adults.

Least Terns in flight show a gray back, narrow wings, and a black leading edge to the primary feathers.

Gull-billed Tern
(*Gelochelidon nilotica*)

CONSERVATION CONCERN SCORE: 13 (Moderate)

OTHER DESIGNATIONS: American Bird Conservancy watch list (Yellow), 2021 USFWS
    Birds of Conservation Concern, State Protected (AL), State Threatened (GA, VA),
    State Endangered (MD), Species of Greatest Conservation Need (FL, MS, NC, SC)

ESTIMATED POPULATION TREND 1966–2019: +24%

SIZE: Length 14 inches; wingspan 34 inches

A male Gull-billed Tern brings a crab to his
mate to strengthen their pair-bond.

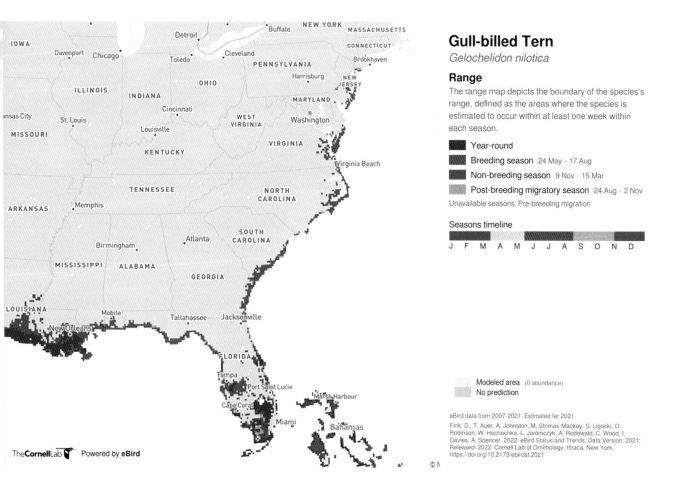

Gull-billed Tern
Gelochelidon nilotica

Range
The range map depicts the boundary of the species's range, defined as the areas where the species is estimated to occur within at least one week within each season.

- Year-round
- Breeding season 24 May - 17 Aug
- Non-breeding season 9 Nov - 15 Mar
- Post-breeding migratory season 24 Aug - 2 Nov

Unavailable seasons: Pre-breeding migration

Seasons timeline

J F M A M J J A S O N D

- Modeled area (0 abundance)
- No prediction

eBird data from 2007-2021. Estimated for 2021.
Fink, D., T. Auer, A. Johnston, M. Strimas-Mackey, S. Ligocki, O. Robinson, W. Hochachka, L. Jaromczyk, A. Rodewald, C. Wood, I. Davies, A. Spencer. 2022. eBird Status and Trends, Data Version: 2021; Released: 2022. Cornell Lab of Ornithology, Ithaca, New York. https://doi.org/10.2173/ebirdst.2021

TheCornellLab   Powered by **eBird**

Species Account Loosely translated, the Gull-billed Tern's scientific name means "swallow of the Nile." This seems appropriate for two reasons: the species can be found along the Nile River, as well as in other warm climates near the seacoast or freshwater, and it has a graceful flight like a swallow. Unlike most other terns, the Gull-billed Tern usually does not capture fish by diving headlong into the water. Instead, it floats gracefully through the air, capturing large insects in flight, or it swoops to the water's surface to take other prey. It has an exceptionally heavy bill for a tern, allowing this species to consume a variety of foods, including insects, crabs, earthworms, frogs, lizards, shrimp, fish, and the chicks of other birds. In some places, this species has become such an effective predator of Least Tern chicks that biologists have resorted to addling (terminating the development of) the eggs of Gull-billed Terns to lessen the impact on Least Terns' reproduction success.

Colonial nesters, Gull-billed Terns typically choose nesting locations in coastal areas, such as on dunes or barrier islands. They often nest in close proximity to Black Skimmers and Least Terns, and these birds may benefit from the Gull-billed Terns' aggressive defensive flights to discourage predators from entering the nesting colony. In recent years, the birds appear to be nesting more often in coastal marshes, as these habitats are visited less frequently by people. Human disturbance can cause adults to leave the nest, exposing the eggs to predators. When humans come too close to a nesting colony, chicks may try to swim to safety, again exposing them to predators or the risk of drowning. As pressures increase in traditional nesting areas, artificially created dredge spoil habitats are being utilized; in some cases, the roofs of buildings have been used for nesting locations. In Virginia, fences have been erected to control mammalian predators, and foxes and raccoons have been removed in an attempt to boost the birds' breeding success.

Large-billed, thick-winged, pale, and long-legged compared to other terns, the Gull-billed Tern has a global distribution that includes six continents. Despite its wide range, this species is rarely seen in large numbers. In the United States, it is found in only a couple of places in Southern California, along the Atlantic coast from New Jersey south, and throughout the Gulf coast. Breeding Bird Survey data indicate a modest increase in the US population since the 1960s, but it is difficult to determine accurate population trends for this species. Some studies suggest a decline in numbers along the Atlantic coast. According to the best current estimates, there are fewer than forty-five hundred breeding pairs in the United States. The highest numbers are in Texas and along the Atlantic coast, where breeding colonies seem to be centered around Virginia and North and South Carolina. It is estimated that there are fewer than forty breeding colonies along the Atlantic, totaling fewer than one thousand pairs.

Identification Gull-billed Terns have heavy black bills and pale bodies and wings. Their legs are black and long for a tern, and their wings are quite wide, especially at the base. In breeding plumage, Gull-billed Terns have solid black caps; these caps nearly disappear during the nonbreeding season, when just a faint smudge remains behind the eye.

Vocalizations Vocalizations consist of various rattles and cackles. At times, they give a three-part call that is similar to the sound made by the katydid (a type of grasshopper).

Nesting Unlike most other tern species, Gull-billed Terns often scrape a nest in the sand and then line its rim with shells or plant material, which may help protect the eggs from being buried by blowing sand. Eggs are light buff with darker speckles, and two or three are typically laid. Young may remain with the parents for three months.

In flight, the Gull-billed Tern appears broad winged. It has a dark bill and dark feet, and the upper wing is almost completely gray and unmarked, setting it apart from other tern species.

Common Tern
(*Sterna hirundo*)

CONSERVATION CONCERN SCORE: 12 (Moderate)

OTHER DESIGNATIONS: 2021 USFWS Birds of Conservation Concern, State Protected (AL), State Special Concern (NC), State Endangered (MD), Species of Greatest Conservation Need (KY, PA, SC, VA)

ESTIMATED POPULATION TREND 1966–2019: −35%

SIZE: Length 12 inches; wingspan 30 inches

The Common Tern is graceful and buoyant in flight. During the breeding season, the bill is blood red with a black tip, the feet are red, and the cap is black. The wings reveal some black on the upper and lower surfaces toward the tips, and the tail is long and forked.

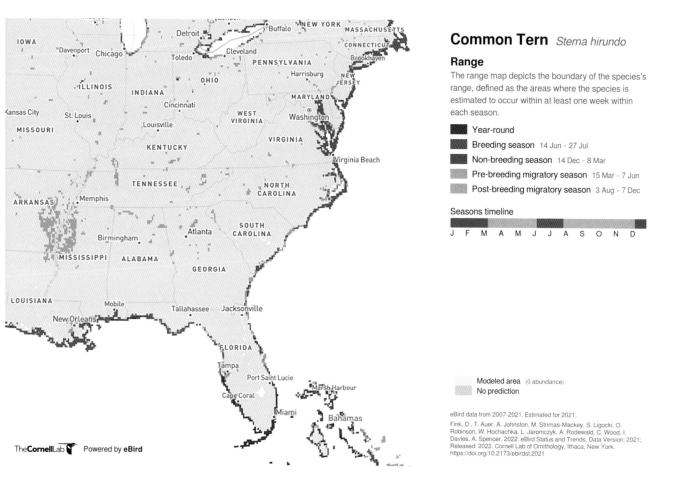

Common Tern *Sterna hirundo*

Range

The range map depicts the boundary of the species's range, defined as the areas where the species is estimated to occur within at least one week within each season.

- Year-round
- Breeding season 14 Jun - 27 Jul
- Non-breeding season 14 Dec - 8 Mar
- Pre-breeding migratory season 15 Mar - 7 Jun
- Post-breeding migratory season 3 Aug - 7 Dec

Seasons timeline

J F M A M J J A S O N D

- Modeled area (0 abundance)
- No prediction

eBird data from 2007-2021. Estimated for 2021.
Fink, D., T. Auer, A. Johnston, M. Strimas-Mackey, S. Ligocki, O. Robinson, W. Hochachka, L. Jaromczyk, A. Rodewald, C. Wood, I. Davies, A. Spencer. 2022. eBird Status and Trends, Data Version: 2021; Released: 2022. Cornell Lab of Ornithology, Ithaca, New York. https://doi.org/10.2173/ebirdst.2021

The Cornell Lab   Powered by **eBird**

Species Account Common Terns are buoyant and agile, making their flight seem effortless. Capable of traveling vast distances, the birds that nest along the Atlantic coastline and in inland regions of the United States and Canada congregate at Cape Cod and other favored spots along the coast before migrating south through the Caribbean in the fall. Eventually, most birds from North America spend their winters along the Atlantic coast in South America, as far south as Argentina, or along the Pacific coast of Central America. Spring migration back to the United States happens remarkably quickly. Birds were tracked with satellites using geolocators, and they made the trip from Venezuela to their breeding grounds in as little as seven days (Nisbet et al. 2011). This species can reach sustained flight speeds of twenty-five to thirty-five miles per hour while migrating. In rare cases, birds banded in North America have crossed the Atlantic Ocean and were recovered in parts of Europe and Africa. There are three subspecies with a near global distribution. Across inland portions of the Appalachians and southeastern United States, the Common Tern is an uncommon migrant. Most sightings occur from mid-April to mid-May and in late July and early October.

Common Terns fly above the water and look down as they search for small fish. When a bird spots its prey, it briefly hovers three to eighteen feet above the water's surface before diving in and grabbing the fish with its bill. A successful bird either turns the fish lengthwise in its bill and eats it on the wing or carries it back to the nest to feed the young. During nesting season, most foraging takes place within a mile or two of the nest site, but some foraging flights may extend a dozen miles or more. Although most terns are solitary hunters, a Common Tern can spot another bird diving for fish two-thirds of a mile away and be drawn to that area. As a result, feeding flocks of hundreds or even a thousand Common Terns can form where prey is abundant, especially if predatory fish are forcing large numbers of small fish to the surface. The most common prey species are the American sand lance, Atlantic herring, pollock, and stickleback. In freshwater locations, common prey species include shiners, perch, and smelt. Although their diet consists mostly of fish, Common Terns also take some shrimp, as well as insects they either capture on the wing or pluck from the water's surface.

During the eighteenth and nineteenth centuries, Common Tern eggs were collected for food along the Atlantic coast, which no doubt reduced their numbers. The birds were later hunted for their feathers, and the Common Tern population on the Atlantic crashed during the 1870s–1890s. In response to the widespread slaughter of terns, herons, and other waterbirds, the Common Tern became a symbol for early conservation efforts that resulted in passage of the Migratory Bird Treaty Act of 1918, which helped many species rebound. Although Common Terns are still considered food in many countries, their population seems to have stabilized. However, declines have been reported in the US population since the 1960s. This is likely due to a combination of factors, including pollution (DDE, PCBs, mercury, lead, and other metals), a growing population of large gulls that feed on tern eggs and young, and human development and recreation in beach habitats.

Identification The Common Tern is one of four similar species that breed in the United States, the others being the Roseate Tern, Forster's Tern, and Arctic Tern. During the breeding season, the Common Tern has a blood-red bill with a black tip, red feet and legs, a gray body with a darker gray back, and a jet-black cap on its head. The tail is long and forked, and the upper wings have a dark leading edge toward the tip. In nonbreeding plumage, the legs, feet, and bill darken toward black, and the front of the black cap whitens.

Vocalizations Common Terns give a harsh, grating two-syllable *kee-arr* call, with a lower-pitched second syllable. They also make short *kip* calls and a rolling *churr* near the nest.

Nesting Once the male establishes his territory, the male and female make several scrapes on the ground in loose substrate with sparse vegetation before settling on a final nest location. A study in North Carolina found that Common Terns prefer nesting locations with 10 to 30 percent cover, which can provide shade or hiding areas for young birds (Soots and Parnell 1975). Nests in coastal areas are often placed in wrack—the debris that accumulates at the high-tide line. Common Terns add material to the nest throughout incubation, especially during an unusually high tide or a flood. By adding nesting material, the birds can raise the eggs as much as an inch or two in a two-hour period. Common Terns typically lay three gray-brown eggs with heavy, dark-brown blotches. The first egg laid is the biggest, the second is slightly smaller, and the third is noticeably smaller.

Common Terns of varying ages transitioning from breeding to nonbreeding plumage.

Black Skimmer
(*Rynchops niger*)

CONSERVATION CONCERN SCORE: 13 (Moderate)

OTHER DESIGNATIONS: American Bird Conservancy watch list (Yellow),
2021 USFWS Birds of Conservation Concern, 2021 Audubon Priority Birds list,
State Special Concern (FL, GA, NC), State Endangered (MD),
Species of Greatest Conservation Need (MS, SC, VA)

ESTIMATED POPULATION TREND 1966–2019: −53%

SIZE: Length 18 inches; wingspan 44 inches

A Black Skimmer prepares to land among a large flock.

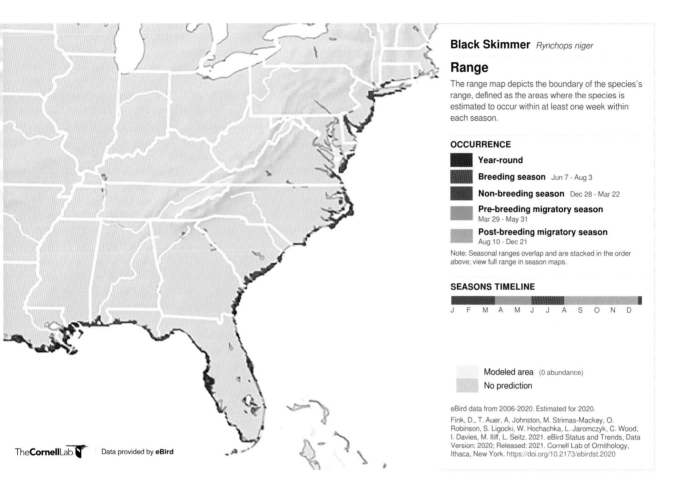

Black Skimmer *Rynchops niger*

Range

The range map depicts the boundary of the species's range, defined as the areas where the species is estimated to occur within at least one week within each season.

OCCURRENCE

Year-round

Breeding season   Jun 7 - Aug 3

Non-breeding season   Dec 28 - Mar 22

Pre-breeding migratory season
Mar 29 - May 31

Post-breeding migratory season
Aug 10 - Dec 21

Note: Seasonal ranges overlap and are stacked in the order above; view full range in season maps.

SEASONS TIMELINE

J F M A M J J A S O N D

Modeled area   (0 abundance)
No prediction

eBird data from 2006-2020. Estimated for 2020.

Fink, D., T. Auer, A. Johnston, M. Strimas-Mackey, O. Robinson, S. Ligocki, W. Hochachka, L. Jaromczyk, C. Wood, I. Davies, M. Iliff, L. Seitz. 2021. eBird Status and Trends, Data Version: 2020; Released: 2021. Cornell Lab of Ornithology, Ithaca, New York. https://doi.org/10.2173/ebirdst.2020

TheCornellLab   Data provided by **eBird**

Species Account Black Skimmers are truly unique. In buoyant flight, they patrol the shallow coastal waters. Flying extremely low, they course back and forth, head down, lower mandible extended just below the surface as they trace a trail in the water, using their sense of touch to detect prey and quickly snap it up in their brightly colored beaks. They resemble a dog with its nose to the ground, tracking an interesting scent on a winding path through the woods. This unusual behavior has led to many of their colorful nicknames, such as "Shearwater," "Sea Dog," "Cutwater," and "Cortaguas," as they are known in Spanish. The birds that breed in Brazil and throughout much of the South American continent are actually two different races of Black Skimmer that nest along interior rivers instead of coastal waters. Globally, there are two other skimmer species. The Indian Skimmer ranges across Southeast Asia, and the African Skimmer is found along rivers and lakes in sub-Saharan Africa and occasionally along the banks of the Nile in Egypt.

Because of their unique feeding behavior, Black Skimmers often hunt within several feet of land, as the shallow water forces small fish (commonly killifish, herring, and

Black Skimmers use their sense of touch to locate fish as they fly through the air while trailing the lower mandible in the water. When the bill touches a fish, this triggers the upper mandible to snap shut and capture the prey.

pipefish) to swim near the surface. Skimmers also preferentially search for fish when the wind speed is relatively low, which results in calmer water. Since they are tactile hunters, skimmers are capable of hunting at all hours of the night and day, whenever conditions are best.

Black Skimmers nest in large colonies, sometimes with other species such as terns, whose aggressive colony defense may protect skimmer nests as well. Despite their short legs, skimmers are fairly agile on the ground. If a head toss or other threat display fails to discourage an intruder, skimmers walk or run after the interloper and chase it away from the nest. The sand can get quite hot during nesting season, and skimmers have been observed diving into the water and wetting their feet and bellies before returning to the nest in an apparent attempt to keep themselves and their eggs or nestlings cool. Adult Black Skimmers occasionally lie flat on the sand, either to regulate their body temperature or to rest their neck muscles, which may be strained by their heavy beaks.

Significant declines in Black Skimmer populations took place in North America during the 1800s, when adults were shot and eggs were taken commercially as a source of food. People in the Amazon basin still take skimmer eggs for food, which may have an impact on populations. In North America, disturbances during the breeding season from beachgoers and off-road vehicles can have a significant negative impact on breeding success. Climate change, pollution, and development of beach habitat are ongoing threats as well. Despite some population increases in California, skimmer populations across North America have continued to decline significantly over the last five decades.

Identification Black Skimmers have almost entirely black upper parts, with pure white bellies and throats and a white tail. The beak is large, with an orange base and a black tip. The lower mandible is significantly longer than the upper mandible. Males are more than one-third heavier than females and larger overall. Juvenile birds resemble adults but are more brownish than black.

Vocalizations Vocalizations include short, nasal barking calls. Near the nest, Black Skimmers give rolling churrs that drop in pitch.

Nesting Courtship can include chasing flights by a pair around the colony or sometimes multiple males chasing a female. The nest is a shallow scrape on the ground in loose sand or shells. Typically, four or five off-white to bluish-green eggs are laid. Both sexes share incubation, which lasts twenty-one to twenty-three days. Initially, the mandibles of young birds are the same length, which allows them to pick up food the parents drop on the ground. When threatened, young birds may lie flat on the sand to avoid detection.

Wood Stork
(*Mycteria americana*)

CONSERVATION CONCERN SCORE: 13 (Moderate)

OTHER DESIGNATIONS: 2021 Audubon Priority Birds list, State Protected (AL), State Special Concern (GA), State Endangered (MS), Species of Greatest Conservation Need (FL, NC), Federally Threatened

ESTIMATED POPULATION TREND 1966–2019: +19%

SIZE: Length 40 inches; wingspan 61 inches

Adult Wood Storks have featherless heads and necks and heavy, powerful bills.

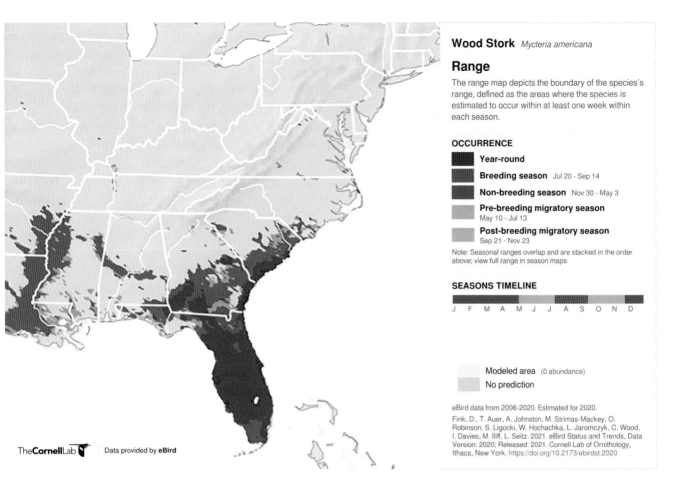

Wood Stork *Mycteria americana*

Range

The range map depicts the boundary of the species's range, defined as the areas where the species is estimated to occur within at least one week within each season.

OCCURRENCE

- Year-round
- Breeding season   Jul 20 - Sep 14
- Non-breeding season   Nov 30 - May 3
- Pre-breeding migratory season
  May 10 - Jul 13
- Post-breeding migratory season
  Sep 21 - Nov 23

Note: Seasonal ranges overlap and are stacked in the order above; view full range in season maps.

SEASONS TIMELINE

J F M A M J J A S O N D

Modeled area (0 abundance)
No prediction

eBird data from 2006-2020. Estimated for 2020.

Fink, D., T. Auer, A. Johnston, M. Strimas-Mackey, O. Robinson, S. Ligocki, W. Hochachka, L. Jaromczyk, C. Wood, I. Davies, M. Iliff, L. Seitz. 2021. eBird Status and Trends, Data Version: 2020; Released: 2021. Cornell Lab of Ornithology, Ithaca, New York. https://doi.org/10.2173/ebirdst.2020

TheCornellLab   Data provided by **eBird**

Species Account Although Breeding Bird Survey data indicate an overall increase in population since the mid-1960s, the story of the Wood Stork is a bit more nuanced. As recently as the 1930s, the Wood Stork population across the southeastern United States was estimated at more than 150,000 birds, with the greatest concentrations occurring in the extensive wetlands of the Everglades in south Florida. As wetlands were drained or hydrology was otherwise altered, populations of these large wading birds steadily declined, to the extent that the Wood Stork was listed as a federally endangered species in 1984. Modern water control programs in the Everglades altered the complex cycle of varying water levels the Wood Stork needs to successfully breed and raise young, resulting in the continued decline of the population in south Florida (by 1995, fewer than five hundred pairs remained). Thankfully, these declines were more than offset by increasing Wood Stork populations in northern Florida, Georgia, Mississippi, and the Carolinas, and in June 2014 the species was downgraded from endangered to threatened.

The Wood Stork is the only stork found in North America. Storks and herons are often confused. In general, herons fly with their necks pulled in toward their bodies, while storks fly with their necks outstretched, resembling long-necked American White Pelicans as flocks circle high in the afternoon thermals. Herons sit perfectly motionless until a prey item comes along; then they ambush it with a quick strike. Storks are more active feeders and tend to wade head-down through freshwater wetlands, sweeping their open beaks through the water. Wood Storks use their sense of touch to feel when a fish or other small animal passes through their open beaks. Once prey is sensed, storks can snap their bills shut within twenty-five milliseconds—the fastest reflex response time of any vertebrate.

Wood Storks need abundant food to successfully nest and feed their chicks. It has been estimated that a pair needs 440 pounds of fish during the nesting season. It is easiest for storks to capture abundant prey when water levels drop and their food becomes trapped in increasingly smaller pools of water. Breeding is triggered only when food is abundant, so when the natural seasonal rise and fall cycles of wetlands are altered, an unintended consequence may be the likelihood of Wood Storks forgoing breeding altogether because prey concentrations never reach appropriate levels. However, if water levels drop too low, predators may have access to stork nests built in trees over standing water. In different ways, Wood Storks' survival depends on the natural cycles of high and low water levels through flood and drought seasons, so maintaining these natural flows is incredibly important.

Because the Wood Stork population is so closely tied to a healthy, functioning wetland system, these birds are considered an "indicator species." Wetlands are complex systems, so it is easier to track the population of an individual (indicator) species and use that to measure the health of the entire system. As work is done to restore the Everglades and other wetlands throughout the Southeast, the Wood Stork may provide important insights into how successful these efforts are.

Identification The Wood Stork is a very large, long-legged wading bird found in cypress swamps and other freshwater wetlands. They have mostly bald, vulture-like heads with long, heavy black beaks. Their plumage is mostly white, with black on the wings and tail visible when in flight. Juvenile birds have a yellowish bill rather than black and lighter-colored heads than adults. The Wood Stork's long legs are dark, but the feet are pinkish or red.

Vocalizations Wood Storks are usually silent, but within the nesting colony they may make a croaking or grunting call. The begging calls of young birds in the nest can be quite loud.

Nesting Wood Storks nest in colonies, often in cypress swamps but also in mangroves or in dead trees surrounded by standing water. The female does most of the nest building, although both sexes participate in incubating the three or four white eggs. After hatching, the young are guarded by the parents, as unmated Wood Storks may attack unprotected young in the colony. The adults regurgitate water over the young birds in the nest to keep them cool on hot days. Young birds first leave the nest at about eight weeks of age.

Wood Storks are rarely far from water, although they are occasionally found in drier habitats.

Wood Storks feed by sweeping their powerful bills back and forth through the water until they touch a prey item, which is quickly snapped up. This young bird's bill is more yellow than that of adults, and it has brownish neck feathers, whereas adults' necks are featherless.

Little Blue Heron
(*Egretta caerulea*)

CONSERVATION CONCERN SCORE: 14 (High)

OTHER DESIGNATIONS: American Bird Conservancy watch list (Yellow),
2021 USFWS Birds of Conservation Concern, In Need of Management (TN),
State Special Concern (FL, GA, NC), Species of Greatest Conservation Need
(KY, MD, SC, VA)

ESTIMATED POPULATION TREND 1966–2019: −48%

SIZE: Length 24 inches; wingspan 40 inches

Juvenile Little Blue Herons are completely white. They are
distinguished from egrets at this age primarily by bill and leg
coloration. This bird will acquire full adult plumage in the second
year of its life, sporting a patchy appearance as it transitions.

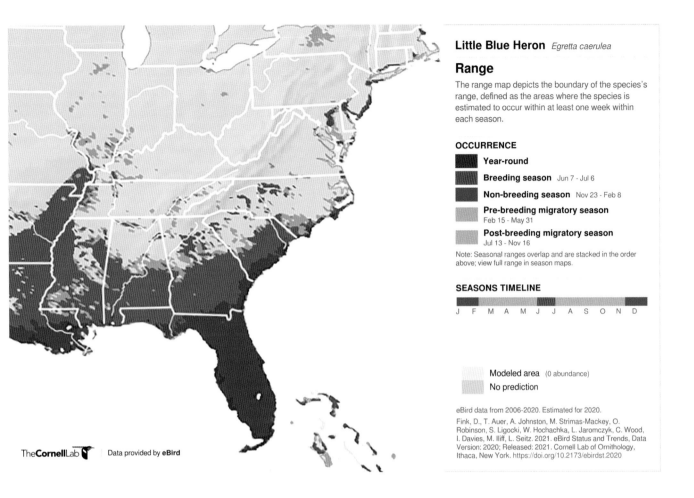

Little Blue Heron *Egretta caerulea*

Range

The range map depicts the boundary of the species's range, defined as the areas where the species is estimated to occur within at least one week within each season.

OCCURRENCE

Year-round

Breeding season Jun 7 - Jul 6

Non-breeding season Nov 23 - Feb 8

Pre-breeding migratory season
Feb 15 - May 31

Post-breeding migratory season
Jul 13 - Nov 16

Note: Seasonal ranges overlap and are stacked in the order above; view full range in season maps.

SEASONS TIMELINE

J F M A M J J A S O N D

Modeled area (0 abundance)
No prediction

eBird data from 2006-2020. Estimated for 2020.

Fink, D., T. Auer, A. Johnston, M. Strimas-Mackey, O. Robinson, S. Ligocki, W. Hochachka, L. Jaromczyk, C. Wood, I. Davies, M. Iliff, L. Seitz. 2021. eBird Status and Trends, Data Version: 2020; Released: 2021. Cornell Lab of Ornithology, Ithaca, New York. https://doi.org/10.2173/ebirdst.2020

TheCornellLab Data provided by eBird

Species Account The Little Blue Heron is a small to medium-sized heron found in a variety of freshwater, marine, and estuarine habitats, including lagoons, tidal flats, rivers, lakes, drainage ditches, flooded fields, and swamps. Fairly common across much of the Southeast, Little Blue Herons can also be found throughout Mexico, Central America, and parts of South America.

Unlike other herons in the region, the immature Little Blue Heron looks completely different from the adult. Immature birds are completely white, while adults have a steely-blue body and a purplish head and neck. During their first summer, young birds undergo a molt, and as they transition from juvenile to adult plumage, their patchy appearance gives rise to names such as Calico, Pied, or Piebald Heron.

As might be expected, given the wide variety of habitats they occupy, Little Blue Herons have a diverse diet that varies based on geography and season; it can also vary as water depth changes locally. Small fish such as anchovies, killifish, gobies, sunfish, drum, and a variety of minnows make up a significant part of the diet. Amphibians, insects, and invertebrates are also taken frequently. Crayfish and grasshoppers seem to be especially common prey items. Unlike some other species of herons and egrets, Little Blue Herons are methodical hunters, standing with the neck fully extended as they scan the water below them. One study from Florida reported that these birds spend 73 percent of their foraging time walking slowly or standing still, waiting for food species to come to them; however, they occasionally employ more active methods such as using their feet to stir or scare potential prey (Rodgers 1983). They forage alone, with other Little Blue Herons, or as part of a group of mixed species that often includes Snowy Egrets or White Ibis.

Compared with the white-plumed egrets, Little Blue Herons were not as heavily targeted by hunters for the millinery (hat-making) trade in the early 1900s; however, they still suffered declines during that time. Today, they occasionally cause problems at aquaculture sites and are sometimes shot, legally or illegally, at these locations. Loss of habitat is likely the primary reason for their ongoing decline, but heavy metals and other pollutants ingested in the foods they eat might play a role. Pesticides, cadmium, lead, and mercury have all been found in body tissue samples and eggs of Little Blue Herons, and studies show slower growth rates and higher mortality in chicks exposed to these contaminants (Spahn and Sherry 1999). Human disturbance at nesting colonies is another factor that occasionally causes nest abandonment. According to Breeding Bird Survey data, declines are most rapid across Alabama, South Carolina, and Florida, while Kentucky and North Carolina have shown slight increases.

Identification Adult birds are steel gray or slate blue on the wings and body, with a deep maroon or purplish neck and head. The beak is bluish at the base, with a dark tip. The legs can appear dark gray or dull greenish. Both sexes are similar. Young birds are completely white, with the bicolored bill of adults. Some birds retain patches of white into their second year before attaining full adult plumage. Immature birds can be distinguished from the similarly sized Snowy Egret because Little Blue Herons lack the bright yellow feet and yellow base of the bill.

Vocalizations Like many wading birds, vocalizations are limited to coarse, guttural croaking.

Nesting Little Blue Herons nest in colonies. They sometimes form their own colonies, but they can also be found in mixed colonies with other species of herons and egrets. When they join a colony of mixed species, the nests of Little Blue Herons are often located along the edges of the colony. The male usually brings sticks to the female, who constructs the nest over a three- to five-day period. Nests are generally placed in a bush or shrub within ten feet of the ground, but they may be as high as thirty feet or more. Typically, three to five pale blue-green eggs are laid, and hatching occurs after twenty to twenty-three days. Incubation is shared by both sexes.

Little Blue Herons spend extended periods seemingly frozen in place as they patiently wait for prey to approach within striking distance.

Reddish Egret
(*Egretta rufescens*)

CONSERVATION CONCERN SCORE: 15 (High)

OTHER DESIGNATIONS: American Bird Conservancy watch list (Red),
2021 USFWS Birds of Conservation Concern, 2021 Audubon Priority Birds list,
State Protected (AL), State Special Concern (FL), Species of Greatest Conservation
Need (MS, SC)

ESTIMATED POPULATION TREND 1966–2019: Not available

SIZE: Length 30 inches; wingspan 46 inches

This Reddish Egret was loping through the shallow water
when it suddenly stopped and threw its wings forward like an
umbrella. The shade provided by the wings may reduce the glare
on the water, helping the birds see their prey more clearly.

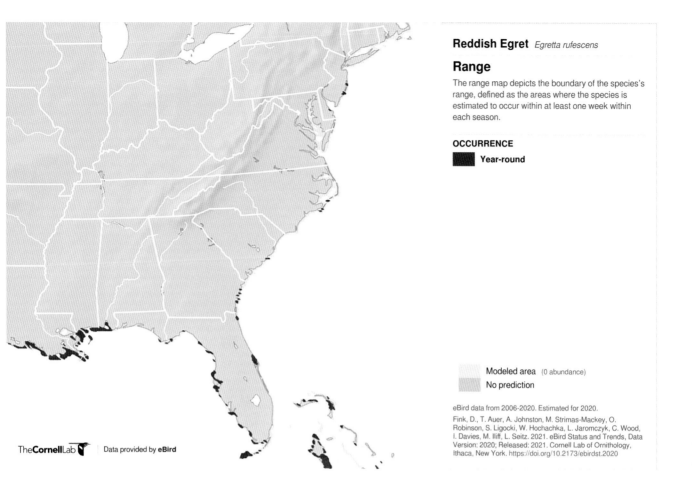

Reddish Egret *Egretta rufescens*

Range

The range map depicts the boundary of the species's range, defined as the areas where the species is estimated to occur within at least one week within each season.

OCCURRENCE

■ Year-round

Modeled area (0 abundance)

No prediction

eBird data from 2006-2020. Estimated for 2020.

Fink, D., T. Auer, A. Johnston, M. Strimas-Mackey, O. Robinson, S. Ligocki, W. Hochachka, L. Jaromczyk, C. Wood, I. Davies, M. Iliff, L. Seitz. 2021. eBird Status and Trends, Data Version: 2020; Released: 2021. Cornell Lab of Ornithology, Ithaca, New York. https://doi.org/10.2173/ebirdst.2020

TheCornellLab | Data provided by **eBird**

Species Account   Like a leaf being blown this way and that by the fickle autumn breeze, the Reddish Egret dances in saltwater shallows, first spinning right, then throwing out a wing and taking two steps to the left before hopping into the air, quickly landing, sprinting forward, and striking with its dagger-like beak to capture a shining minnow that flaps and struggles in an unsuccessful attempt to free itself from the egret's vise-like hold. These master anglers of the shallow saltwater pools of Florida and the other Gulf coast states take an active, almost bizarre approach to hunting compared with other, more stately egrets and herons that sit perfectly still for many minutes before suddenly striking at prey that happens to swim by. Although they sometimes adopt a more sedentary hunting style, it is common to see Reddish Egrets dancing or spinning, often holding out one or both wings to shade the water as they search for food. Using their wings as shade may attract fish looking for shelter, or it may improve the birds' ability to see into the water without the glare of the sun. Either way, because of their antics, it is easy to pick out a Reddish Egret among a group of wading birds, even at quite a distance.

Reddish Egrets are also somewhat unusual among birds of the Southeast because they have two different color phases. Dark-phase birds, which are much more common, are slate gray on the back and wings, with dark legs and a rich chestnut color on the neck and head. Light-phase birds are pure white with dark legs. Both phases have a bicolored beak that is pinkish at the base and turns to black at the tip. Color phase is not relevant for breeding purposes, and it is not unusual to have a pair with one light-phase and one dark-phase bird. Chicks may be either light phase or dark phase, regardless of their parents' color phase, and they retain that coloration throughout their lives.

Reddish Egret's favorite prey items include sheepshead minnow, sailfin molly, pinfish, striped mullet, various killifish, and tidewater silverside. Although they also eat shrimp, frogs, and crabs, the bulk of their diet consists of small minnows. They are more closely tied to saltwater than most of the other herons and egrets of the Southeast and are commonly found in shallow bays, lagoons, salt marshes, and shorelines where water depths are three to four inches or less.

Although accurate data are unavailable through the Breeding Bird Survey, it is thought that Reddish Egret populations declined drastically in the 1800s because the birds were hunted for their elaborate feather plumes. Light-phase birds were especially popular with hunters, which may explain why they are much less common today than dark-phase birds. For much of the 1920s and 1930s the birds were reportedly absent from a good portion of their US range. Today, human disturbance and coastal development are probably the biggest threats. Reddish Egrets may still number fifteen thousand or more pairs throughout their global range, which includes coastal areas of Mexico, Central America, and the Caribbean; however, there are likely only about two thousand pairs in the United States.

Identification Reddish Egrets are large wading birds that are either all white (light phase) or slate gray with reddish necks and heads (dark phase) that may appear golden or bronze in early morning or late afternoon light. Long feathers on the head and neck give them a maned look. The beak is black at the tip and pinkish at the base. At the peak of breeding season, turquoise or lavender colors appear between the base of the beak and the eye. Dark-phase birds may be confused with the Little Blue Heron, and light-phase birds may be mistaken for the similar Snowy Egret, which has bright yellow feet.

Vocalizations Like most herons and egrets, Reddish Egrets are limited to coarse croaking or grunting and are usually silent, except around the nest.

Nesting The male and female choose the nest site together. It is usually located in a shrub or mangrove but is sometimes right on the ground. The nests are often built over water and may be part of a mixed colony of other wading birds. The female usually constructs the nest platform, although the male may help by bringing sticks. The female typically lays three or four pale blue-green eggs, and the young hatch after about twenty-five days.

An adult Reddish Egret in breeding plumage dispatches a small minnow it just captured. This individual is a dark-phase bird. Although it has some white feathering, it is not a light phase–dark phase hybrid. The white feathers are caused by leucism, or a lack of melanin pigment.

The long feather plumes on the neck give this Reddish Egret the appearance of having a golden mane in the late-afternoon light.

Yellow-crowned Night-Heron
(*Nyctanassa violacea*)

CONSERVATION CONCERN SCORE: 12 (Moderate)

OTHER DESIGNATIONS: State Special Concern (GA),
    Species of Greatest Conservation Need (KY, MS, NC, PA, SC, VA)

ESTIMATED POPULATION TREND 1966–2019: −10%

SIZE: Length 24 inches; wingspan 42 inches

An adult Yellow-crowned Night-Heron in breeding plumage hunts in a small wetland.

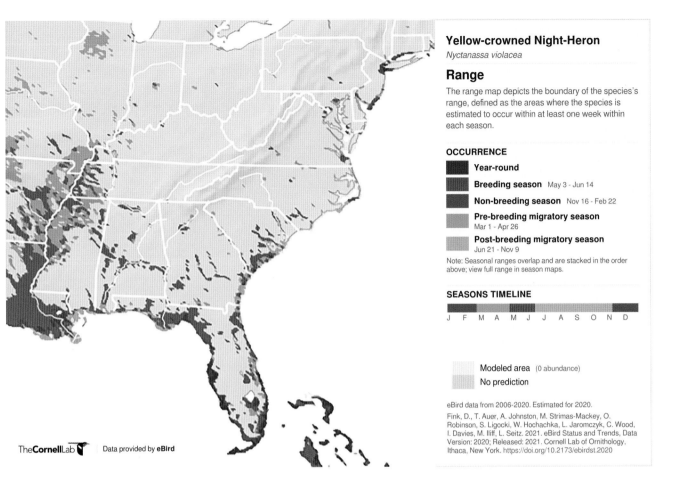

Yellow-crowned Night-Heron
Nyctanassa violacea

Range

The range map depicts the boundary of the species's range, defined as the areas where the species is estimated to occur within at least one week within each season.

OCCURRENCE

- **Year-round**
- **Breeding season** May 3 - Jun 14
- **Non-breeding season** Nov 16 - Feb 22
- **Pre-breeding migratory season** Mar 1 - Apr 26
- **Post-breeding migratory season** Jun 21 - Nov 9

Note: Seasonal ranges overlap and are stacked in the order above; view full range in season maps.

SEASONS TIMELINE

J F M A M J J A S O N D

Modeled area (0 abundance)
No prediction

eBird data from 2006-2020. Estimated for 2020.

Fink, D., T. Auer, A. Johnston, M. Strimas-Mackey, O. Robinson, S. Ligocki, W. Hochachka, L. Jaromczyk, C. Wood, I. Davies, M. Iliff, L. Seitz. 2021. eBird Status and Trends, Data Version: 2020; Released: 2021. Cornell Lab of Ornithology, Ithaca, New York. https://doi.org/10.2173/ebirdst.2020

TheCornellLab | Data provided by **eBird**

Species Account The Yellow-crowned Night-Heron is short and stocky compared with other herons. Its beak is shorter and heavier than that of other wading birds, but it is perfectly suited for grabbing and breaking apart the species' favorite prey: crabs. Although they also consume fish, worms, small mammals, insects, and other food items, the birds' diet consists primarily of a wide variety of crabs, including sand fiddler and mud fiddler crabs, marsh crabs, ghost crabs, mole crabs, common mud crabs, Atlantic blue crabs, lady crabs, rock crabs, and toad crabs. A study that analyzed the stomach contents of more than one hundred individual birds from the southeastern United States found that their diet consisted of crustaceans (79 percent), insects (18 percent), vertebrates (1 percent), gastropods (<1 percent), and arachnids (<1 percent) (Riegner 1982). In inland areas, Yellow-crowned Night-Herons feed almost exclusively on crayfish.

When hunting, the Yellow-crowned Night-Heron is largely sedentary, spending as much as 80 percent of its time standing motionless in the shallow water of tidal marshes, tide pools, mudflats, swamps, or other habitats in search of prey. After a successful strike,

Juvenile Yellow-crowned Night-Herons lack the steel-gray plumage and bold facial patterns of adults. Though they are very similar to young Black-crowned Night-Herons, they have a heavier, darker bill and small whitish spots rather than streaks on the neck.

the birds usually eat small crabs whole; they remove the legs of larger crabs before consuming them. Tide levels seem to be more important than time of day for feeding purposes. The birds feed most actively within three hours of low tide, regardless of whether it is night or day. Although a number of Yellow-crowned Night-Herons may use the same habitat for feeding, they typically do not tolerate any other individuals of their species within fifteen feet when hunting. According to a study in Virginia, vegetation type, wetland shape (preferably long and linear, with more edge habitat), and distance from the nest were important factors in determining preferred foraging locations (Bentley 1994). Although most common along the Atlantic and Gulf coasts, Yellow-crowned Night-Herons can also be found as far north as Ohio, Indiana, and Minnesota during the breeding season.

The overall population size and population trends of this species are difficult to measure. Although one hundred pairs or more may nest in mixed breeding colonies with other wading bird species, Yellow-crowned Night-Herons may also nest singly or with a handful of other pairs of their own species in small, isolated breeding groups. Because of this tendency, and because the birds' dark plumage makes them difficult to pick up during aerial surveys, this is one of the more difficult heron species to monitor. However, it appears that the population grew fairly dramatically from the 1920s through the 1950s, with new breeding records from nearly a dozen states. Since 1960, the best information indicates a slight decline in the North American population, with the loss of wetland habitat the most likely cause.

Identification Adult Yellow-crowned Night-Herons have bold facial patterns of alternating black and cream patches, with white plumes at the back of the head. They have a bright red iris and a heavy black beak. The neck and belly are steel gray, with darker streaking on a gray background on the wings and back. The legs are bright yellow. Juvenile birds are brown overall, with white spots and streaks; their legs are greenish, and the iris is golden orange. Young birds are very similar in appearance to juvenile Black-crowned Night-Herons.

Vocalizations Yellow-crowned Night-Herons give a harsh, guttural *quawk* or *skow* call.

Nesting The Yellow-crowned Night-Heron may be part of a larger nesting colony of other wading birds, or it may nest individually or as part of a small group. In Virginia, the average colony size was four nesting pairs, with 25 percent of nests occurring singly. Oak and loblolly pine are the most common tree species used for nesting. Nests are commonly placed six to fifty-four feet above the ground, often in the last fork of a branch, possibly to reduce predation from raccoons and opossums. Three to six pale bluish-green eggs are laid. Birds that nest in subdivisions in the Southeast have been involved in problematic interactions with people.

Swallow-tailed Kite
(*Elanoides forficatus*)

CONSERVATION CONCERN SCORE: 12 (Moderate)

OTHER DESIGNATIONS: American Bird Conservancy watch list (Yellow), 2021 USFWS Birds of Conservation Concern, State Protected (AL), State Special Concern (GA), State Endangered (MS), Species of Greatest Conservation Need (FL, KY, NC, SC)

ESTIMATED POPULATION TREND 1966–2019: +96%

SIZE: Length 22 inches; wingspan 51 inches

An adult Swallow-tailed Kite carries Spanish moss to use in nest construction.

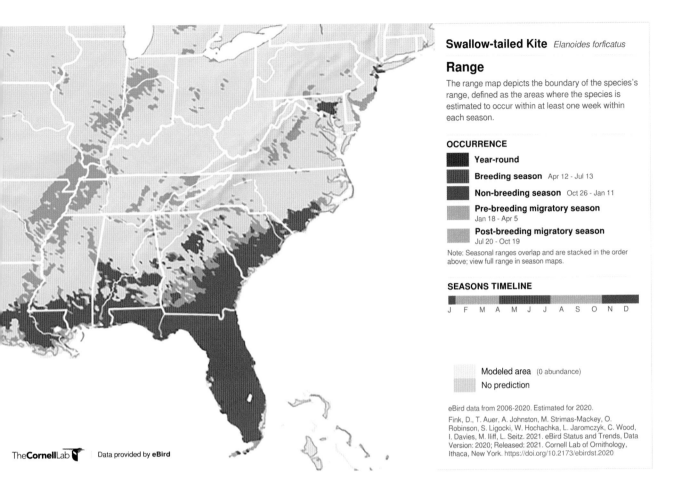

Swallow-tailed Kite *Elanoides forficatus*

Range

The range map depicts the boundary of the species's range, defined as the areas where the species is estimated to occur within at least one week within each season.

OCCURRENCE

- **Year-round**
- **Breeding season** Apr 12 - Jul 13
- **Non-breeding season** Oct 26 - Jan 11
- **Pre-breeding migratory season** Jan 18 - Apr 5
- **Post-breeding migratory season** Jul 20 - Oct 19

Note: Seasonal ranges overlap and are stacked in the order above; view full range in season maps.

SEASONS TIMELINE

J F M A M J J A S O N D

- Modeled area (0 abundance)
- No prediction

eBird data from 2006-2020. Estimated for 2020.

Fink, D., T. Auer, A. Johnston, M. Strimas-Mackey, O. Robinson, S. Ligocki, W. Hochachka, L. Jaromczyk, C. Wood, I. Davies, M. Iliff, L. Seitz. 2021. eBird Status and Trends, Data Version: 2020; Released: 2021. Cornell Lab of Ornithology, Ithaca, New York. https://doi.org/10.2173/ebirdst.2020

TheCornellLab | Data provided by **eBird**

Species Account Incredibly agile and graceful in flight, Swallow-tailed Kites are almost unmistakable with their long, forked tails and striking black-and-white coloration. They seem to float through the sky, using their lengthy tails to make quick, twisting turns to catch flying insects on the wing or scanning the treetops and grabbing an unsuspecting lizard from among the leaves. The birds capture all their prey with their feet, and they often eat in flight by using their beaks to tear bites from the food clutched in their talons. The bulk of the diet is made up of insects, with grasshoppers being a favorite. Stinging insects such as wasps, bees, and hornets also account for a significant portion of the diet. Swallow-tailed Kites have been known to take whole wasp nests back to their own nests, where they feed on the larvae and sometimes work the wasp nest into their own nest structure. Apparently, the stomach lining of Swallow-tailed Kites is thicker and spongier than that of other birds of prey, allowing them to feed on stinging insects as well as fire ants and other species that most birds would avoid. In addition to insects, kites take nestling birds, frogs, lizards, snakes, bats, and some small fish. In the tropics, the birds also feed on fruit during much of the year. Studies from Florida indicate that the diet may shift based on local weather conditions, with more frogs taken in wet years and a higher percentage of insects in drier years (Meyer and Collopy 1990).

Swallow-tailed Kites prefer large pines or cypress trees for nesting sites. Trees that have an open crown and are taller than surrounding trees are especially favored. Wetlands, prairies with scattered large trees, pine forests with interspersed wetlands, and clusters of larger trees along rivers and sloughs are favored habitats. Australian pine, a nonnative species introduced to Florida, is occasionally used by the kites for nesting. Unfortunately, nests placed in Australian pines tend to fail at significantly higher rates than those placed in native pines or cypress trees.

Although Breeding Bird Survey data indicate a rapidly increasing population in the United States, these data are not considered reliable because of the relatively low number of transects that encounter Swallow-tailed Kites. In fact, this species has experienced a significant contraction of its range in the United States over the last century. Prior to the 1940s, nesting records existed for Arkansas, Oklahoma, Kentucky, Missouri, Kansas, Illinois, Iowa, Wisconsin, and Minnesota; Swallow-tailed Kites were also seen frequently in Indiana, Ohio, Nebraska, and Tennessee. Today the Swallow-tailed Kite has been extirpated from all these states. The current US population is estimated to be fewer than five thousand birds, with roughly 60 percent of the total population occurring in Florida. South Carolina has an estimated 110 breeding pairs, and no other state is believed to have more than 100 breeding pairs. Even across its remaining range in the southeastern United States, some well-documented breeding areas no longer have nesting kites, and the species' geographic range has continued to shrink even in states that still host nesting birds. A separate population of Swallow-tailed Kites nests in portions of Central and South America.

The biggest threats to this species include destruction of remaining wetlands and conversion of habitat to residential, urban, or agricultural uses such as citrus production. Intensive forest production practices often result in solid stands of similarly aged trees, with no pines or cypress trees mature enough to support nesting. In late summer, congregations of more than one thousand Swallow-tailed Kites occur in portions of Florida prior to migration to nonbreeding areas farther south. These large concentrations of birds make them more vulnerable to human disturbance or weather events such as hurricanes. The subspecies that breeds in the southeastern United States has been considered for listing as a federally endangered species.

Identification A large raptor with a four-foot wingspan, the Swallow-tailed Kite has a white head and a contrasting black-and-white pattern underneath, with a long, deeply forked tail. Males and females are identical.

Swallow-tailed Kites frequently snatch prey from the tops of bushes and trees. They consume smaller prey items while in flight.

Vocalizations A shrill, whistled *klee-klee-klee* is the most common vocalization.

Nesting Nests are platforms made of sticks. They are constructed by both sexes. The nests are located in mature trees that are often taller than the surrounding forest, and they are usually placed at least sixty feet above the ground. Two creamy white eggs are typically laid. Young birds remain in the nest for five to six weeks.

The Swallow-tailed Kite's white head stands out from the dark gray body and black shoulders. The long, split tail is spread to make quick course corrections while in flight.

Northern Harrier
(*Circus hudsonius*)

CONSERVATION CONCERN SCORE: 11 (Moderate)

OTHER DESIGNATIONS: 2021 USFWS Birds of Conservation Concern,
2021 Audubon Priority Birds list, State Protected (AL), State Special Concern (MD),
Species of Greatest Conservation Need (KY, NC, PA, TN, VA, WV)

ESTIMATED POPULATION TREND 1966–2019: −35%

SIZE: Length 18 inches; wingspan 43 inches

A female Northern Harrier cruises low over the prairie
grasses searching for prey that is unaware of her presence.
Formerly called the Marsh Hawk, Northern Harriers are
residents of open country such as marshes and grasslands.

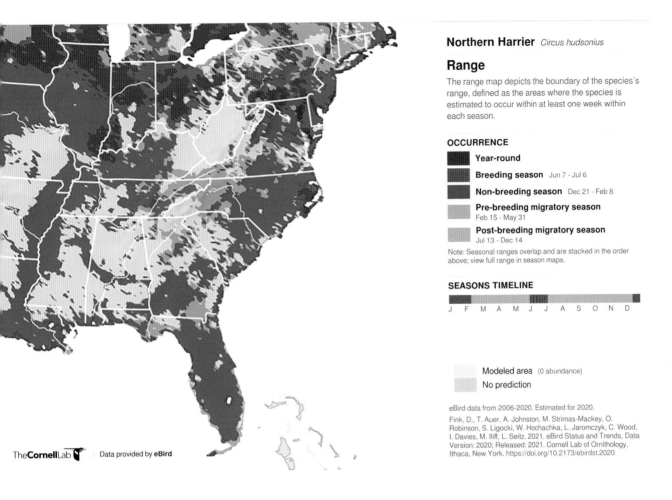

Northern Harrier *Circus hudsonius*

Range

The range map depicts the boundary of the species's range, defined as the areas where the species is estimated to occur within at least one week within each season.

OCCURRENCE

■	**Year-round**
■	**Breeding season** Jun 7 - Jul 6
■	**Non-breeding season** Dec 21 - Feb 8
■	**Pre-breeding migratory season** Feb 15 - May 31
■	**Post-breeding migratory season** Jul 13 - Dec 14

Note: Seasonal ranges overlap and are stacked in the order above; view full range in season maps.

SEASONS TIMELINE

J F M A M J J A S O N D

Modeled area (0 abundance)
No prediction

eBird data from 2006-2020. Estimated for 2020.
Fink, D., T. Auer, A. Johnston, M. Strimas-Mackey, O. Robinson, S. Ligocki, W. Hochachka, L. Jaromczyk, C. Wood, I. Davies, M. Iliff, L. Seitz. 2021. eBird Status and Trends, Data Version: 2020; Released: 2021. Cornell Lab of Ornithology, Ithaca, New York. https://doi.org/10.2173/ebirdst.2020

TheCornellLab | Data provided by **eBird**

Species Account Cruising low over grasslands or open marshes, Northern Harriers fly with powerful wing beats followed by periods of gliding. They hold their wings upward in a V shape as they scan the vegetation below, occasionally hovering in place or whirling back in a tight half circle before dropping to the ground to take their prey. With this unusual flight pattern and their long tails, long, narrow wings, and white rump patch, Northern Harriers are fairly easy to distinguish from other raptors. In fact, they are probably most often confused with the Short-eared Owl, which is found in similar habitats. It is not uncommon for Northern Harriers to patrol a weedy field, wetland, or prairie during the daylight hours, only to be joined by Short-eared Owls as dusk approaches, then yielding the air to the owls as it becomes fully dark.

Northern Harriers have an unusually shaped face. Their facial feathers form a disc or bowl that helps funnel sounds to their ears and improves the sensitivity and directional capability of their hearing, similar to many owls. This allows the birds to use both hearing and vision to locate their quarry. Their diet consists largely of small mammals such as voles, mice, and other rodents. However, they also take a fair number of birds,

Seen from slightly above, the distinctive white rump patch at the base of the long tail of this Northern Harrier is clearly visible. The genus name *Circus* is a reference to the birds' circling flight as they repeatedly cruise the same sections of habitat.

including sparrows and other small species up to the size of doves and small ducks. In addition, Northern Harriers eat large insects, snakes, lizards, and frogs. Male Northern Harriers are more likely to prey on other birds, while females are more likely to take mammals. The female birds exclude the smaller males from preferred feeding areas during the nonbreeding season.

Although Northern Harriers are typically seasonally monogamous, an abundance of prey during nesting season may allow males to mate with more than one female. In the best territories, males may have as many as five mates in one season, nesting in close proximity to one another. The male performs a "sky dance" display by flying in swooping dips over the prospective nesting area. The female follows the male, and the dance ends on the ground at a potential nest location. Once the pair-bond is established, the male often feeds the female by transferring prey to her while both are in flight. The male continues to supply the female with food during incubation and provides food for the young until they leave the nest; the female typically feeds the young herself once they are fledged.

In the Appalachians and the Southeast, the Northern Harrier is seen primarily during the nonbreeding season, although the species nests in parts of coastal North Carolina, Virginia, West Virginia, Maryland, and Pennsylvania. Across its range, the species is still fairly common in most locations. However, according to Breeding Bird Survey data and fall migration counts at Hawk Mountain, Pennsylvania, it is in decline. Northern Harriers have disappeared from some parts of their range due to loss of

Unlike females and juveniles, the male Northern Harrier is light gray, although he too sports the white rump patch visible in flight. Because of their coloration, males are sometimes referred to as gray ghosts.

habitat through the drainage of wetlands, conversion of grasslands, and transition of open, early successional habitats to mature forest.

Identification Northern Harriers are long-tailed, long-winged birds of prey that are often seen flying low to the ground over open grasslands or marshes. The females are significantly larger and heavier than the males and are brown overall, with brown streaks visible on the breast and the underside of the wings. Males have silver-gray backs, with some brown splotches on the chest and belly. Juvenile birds resemble the female but have rich buffy underparts. All birds in all seasons exhibit a large white patch at the base of the tail when seen from above.

Vocalizations Northern Harriers are typically silent except during the nesting season, when they give shrill *kee-kee-kee-kee* or sharp whistled calls during courtship or to sound an alarm.

Nesting Nests are built on the ground, often in a clump of vegetation. They are commonly placed in wet habitats, likely because of reduced predator levels. The male initiates nest construction, but the female completes the nest. Four to six light-blue eggs are typically laid, but larger clutches are possible. Clutch size is thought to be related to the amount of food supplied by the male during courtship and nest construction.

Snail Kite
(*Rostrhamus sociabilis*)

CONSERVATION CONCERN SCORE: Not available. Although the global population is secure, with a score of 9, the US population would score much higher.

OTHER DESIGNATIONS: 2021 Audubon Priority Birds list, Species of Greatest Conservation Need (FL), Federally Endangered

ESTIMATED POPULATION TREND 1966–2019: Not available

SIZE: Length 17 inches; wingspan 42 inches

The curved bill of the Snail Kite is the perfect tool for reaching into a snail shell and extracting a meal.

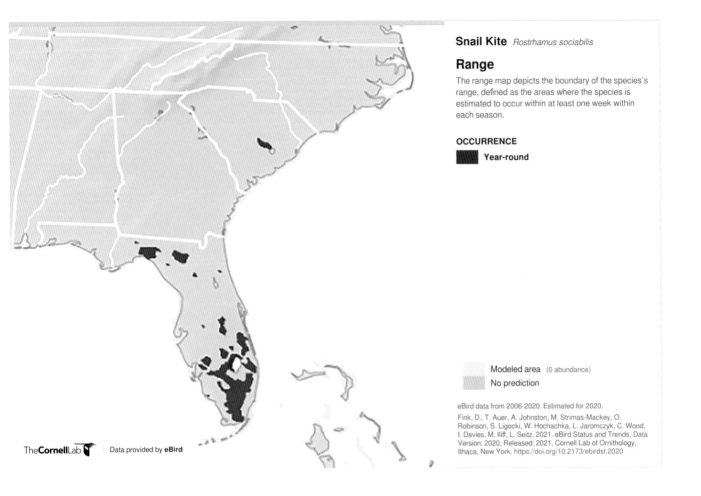

Snail Kite *Rostrhamus sociabilis*

Range

The range map depicts the boundary of the species's range, defined as the areas where the species is estimated to occur within at least one week within each season.

OCCURRENCE

█ Year-round

▨ Modeled area (0 abundance)
▨ No prediction

eBird data from 2006-2020. Estimated for 2020.
Fink, D., T. Auer, A. Johnston, M. Strimas-Mackey, O. Robinson, S. Ligocki, W. Hochachka, L. Jaromczyk, C. Wood, I. Davies, M. Iliff, L. Seitz. 2021. eBird Status and Trends, Data Version: 2020; Released: 2021. Cornell Lab of Ornithology, Ithaca, New York. https://doi.org/10.2173/ebirdst.2020

TheCornellLab | Data provided by **eBird**

Species Account The Snail Kite is perfectly designed to catch and eat one thing—apple snails—and usually ignores other types of prey. Favoring shallow, freshwater wetlands with large areas of clear water unobstructed by vegetation, the Snail Kite cruises low in search of snails, which it can snatch from water up to six inches deep. The bird drops down on its unsuspecting quarry, grabbing it out of the water or off aquatic vegetation with its talons. The bird then carries the snail to a perch, where it uses its long, scythe-like upper mandible to reach in and extract the snail from its shell. Although perfectly capable of capturing its own meals, the Snail Kite is not above trying to steal from Limpkins, which also dine on snails and often occur in the same marshes. Recently, a nonnative snail species has arrived in Florida and appears to be displacing the native apple snail. It is possible that the additional food provided by the larger introduced species could benefit Snail Kite populations, at least in the short term.

Unusual for a bird of prey, the Snail Kite is a colonial nester at times. Colonies vary in size and are often located at the edges of waterbird nesting colonies that include Anhingas, herons, or ibis. These colonies may exist for a number of years before being

The female Snail Kite is streaked with tawny brown,
in contrast to the solid steely gray of the male.

abandoned. In times of drought, it is not uncommon for Snail Kites to forgo breeding altogether for a year or more until conditions improve. However, at times of ideal prey availability, Snail Kites may raise multiple broods in succession during an eight- to ten-month period.

As part of their courtship ritual, male Snail Kites do much of the nest building. They construct a loose platform of sticks and then perform a slow "butterfly" flight near a female or offer her a stick. Once a pair is established, the female completes construction of the nest, which is usually placed low in a tree or shrub or in herbaceous vegetation over water. Nests in herbaceous vegetation have lower success rates, probably because they are more prone to collapse as the nestlings grow; they may also be more vulnerable to predators such as alligators. To increase productivity, biologists have placed pole-mounted baskets in wetlands where there are insufficient trees or shrubs for nesting, and Snail Kites have readily adopted these artificial nest locations. Once the chicks have fledged, either one of the adults may leave and begin a second nest with a different mate if food availability is high.

Because Snail Kites require very specific conditions for successful hunting, they are often forced to wander, looking for new habitat during times of drought or high water. They are also vulnerable to changing water levels due to human development or agricultural needs. During the twentieth century, much of the Everglades in south Florida was drained, leaving few areas suitable for Snail Kites. By 1972, only sixty-five birds remained in the state—the only place they occur in the United States. Although Snail Kites are common across southern Mexico and South America, the US population (formerly known as the Everglades Kite) stands at approximately one thousand birds, all occurring in central and south Florida. The species has been listed as federally endangered since 1967. As efforts are made to restore the natural hydrology in portions of the Everglades, it is hoped that this species will rebound along with the wetland habitat.

Identification Adult males are slaty black, with a white rump patch visible in flight and a white terminal band on the tail. The feet and legs are orange, the beak is orange with a black tip, and the iris is bright red. Females and young birds have brown backs, brown streaks on the chest, and buffy stripes over the eyes. Both sexes have the distinctive hooked beak.

Vocalizations Adult birds give a cackling *ka-ka-ka-ka-ka-ka* call.

Nesting This species typically lays two or three eggs, although smaller clutches may occur in response to poor habitat conditions. The eggs are white with brownish speckles.

Burrowing Owl
(*Athene cunicularia*)

CONSERVATION CONCERN SCORE: 12 (Moderate)

OTHER DESIGNATIONS: 2021 USFWS Birds of Conservation Concern, State Protected (AL), State Special Concern (FL)

ESTIMATED POPULATION TREND 1966–2019: −32%

SIZE: Length 10 inches; wingspan 21 inches

This adult Burrowing Owl was sleepily stretching until it spotted something (an insect?) that brought it to full alert.

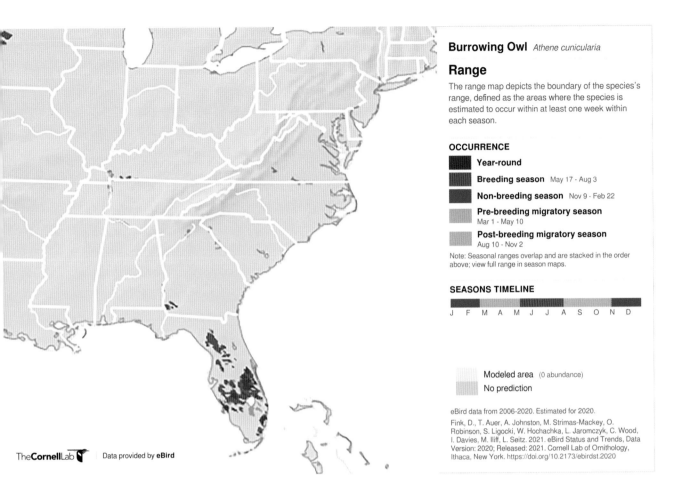

Burrowing Owl *Athene cunicularia*

Range

The range map depicts the boundary of the species's range, defined as the areas where the species is estimated to occur within at least one week within each season.

OCCURRENCE

Year-round

Breeding season   May 17 - Aug 3

Non-breeding season   Nov 9 - Feb 22

Pre-breeding migratory season
Mar 1 - May 10

Post-breeding migratory season
Aug 10 - Nov 2

Note: Seasonal ranges overlap and are stacked in the order above; view full range in season maps.

SEASONS TIMELINE

J F M A M J J A S O N D

Modeled area   (0 abundance)
No prediction

eBird data from 2006-2020. Estimated for 2020.

Fink, D., T. Auer, A. Johnston, M. Strimas-Mackey, O. Robinson, S. Ligocki, W. Hochachka, L. Jaromczyk, C. Wood, I. Davies, M. Iliff, L. Seitz. 2021. eBird Status and Trends, Data Version: 2020; Released: 2021. Cornell Lab of Ornithology, Ithaca, New York. https://doi.org/10.2173/ebirdst.2020

TheCornellLab   Data provided by **eBird**

Species Account The wide-ranging Burrowing Owl can be found in a multitude of habitats, including deserts, grasslands, airports, parks, and suburban areas. There are minor geographic differences in plumage, and several different subspecies are found across North America, in Central America, and as far south as the tip of South America. Because these owls nest underground, they prefer open land with only a few scattered perch locations for lookouts. Burrowing Owls are often found where other animals have excavated burrows the owls can utilize as nest cavities. Suitable owl burrows include those dug by prairie dogs, tortoises, badgers, skunks, and armadillos, among other species. As long as the soil is loose, Burrowing Owls can dig their own burrows using their beaks to scrape and their feet to kick dirt out of the hole, but the process usually takes several days. One or more satellite burrows are used for roosting by the male and often by the young owls once they leave the nest. Owls may also use other abandoned burrows or clumps of grass to roost in.

As one might expect, given their wide geographic range, Burrowing Owls feed on an array of prey items. Insects make up the bulk of the diet, with grasshoppers, crickets,

A Burrowing Owl emerges from its burrow as the sun sinks toward the horizon and the light of day fades.

moths, dragonflies, and beetles among the most common. Lizards, snakes, rodents, bats, tortoises, and other birds are also taken. Interestingly, male and female Burrowing Owls tend to specialize in different types of prey, with females taking insects during daytime hunting and males taking vertebrates that are usually captured at night. The birds may hunt at any time of the night or day and are capable of flying and hovering to search for prey. At times they may be seen walking or running on the ground to capture a meal.

Relatively tame and trusting, Burrowing Owls have been known by many nicknames over the years. Cowboys in the West called them "Howdy Birds" because of the way they rapidly bowed their heads in agitation when humans approached their burrows. In other places they have been known as Prairie Dog Owls, Ground Owls, and Billy Owls.

In the United States, the Burrowing Owl is largely a western bird. It is commonly found in the short grass prairies east of the Rocky Mountains and in the deserts of the Southwest. Individual birds occasionally show up in much of the East after the breeding season, but the only year-round population east of the Mississippi River occurs in Florida. Although overall population numbers show only a modest decline since the mid-1960s, the Florida population is in serious decline, experiencing an estimated 90 percent drop according to Breeding Bird Survey data. Loss of habitat to agriculture and development, collisions with vehicles, and fatal interactions with domestic cats and dogs have contributed to this decline. The Burrowing Owl is endangered in Canada and listed as a Species of Special Protection in Mexico.

Identification   The Burrowing Owl is a small, long-legged owl with no ear tufts. Burrowing Owls are brown with white spots on the head and back and have a white belly with brown horizontal barring. They have a white throat patch and bright yellow eyes.

Vocalizations   The owls give a cooing call somewhat like that of a dove. They also give a chattering call when alarmed and can make a rattling noise similar to a rattlesnake when disturbed in their burrows.

Nesting   Depending on whether the birds dig their own burrow or take over another animal's abandoned burrow, the nest site is usually no more than three feet underground at the end of a tunnel that can measure nine feet or longer. The entrance is usually four to six inches in diameter. The nest itself is often lined with manure, feathers, or grass. Anywhere from three to a dozen white eggs are laid, and incubation lasts about twenty-eight days. The male brings food to the nest for the female and the chicks until they reach one to two weeks of age. At that point, both parents begin hunting to supply food for the young birds' growing appetites.

Red-headed Woodpecker
(*Melanerpes erythrocephalus*)

CONSERVATION CONCERN SCORE: 13 (Moderate)

OTHER DESIGNATIONS: American Bird Conservancy watch list (Yellow), 2021 USFWS Birds of Conservation Concern, 2016 Partners in Flight Species of Continental Concern (Yellow), Species of Greatest Conservation Need (FL, KY, MD, MS, NC, PA, SC, TN, WV)

ESTIMATED POPULATION TREND 1966–2019: −53%

SIZE: Length 9 inches; wingspan 17 inches

The adult Red-headed Woodpecker is unmistakable with its completely red head, white belly, black back, and large white wing patches.

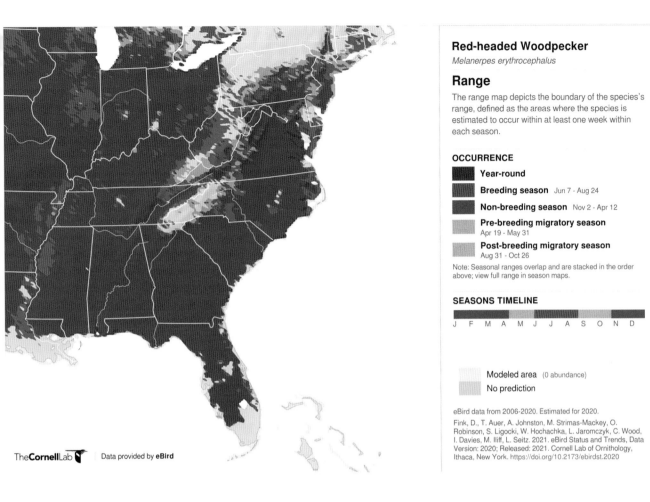

Red-headed Woodpecker
Melanerpes erythrocephalus

Range

The range map depicts the boundary of the species's range, defined as the areas where the species is estimated to occur within at least one week within each season.

OCCURRENCE

Year-round

Breeding season   Jun 7 - Aug 24

Non-breeding season   Nov 2 - Apr 12

Pre-breeding migratory season
Apr 19 - May 31

Post-breeding migratory season
Aug 31 - Oct 26

Note: Seasonal ranges overlap and are stacked in the order above; view full range in season maps.

SEASONS TIMELINE

J F M A M J J A S O N D

Modeled area   (0 abundance)
No prediction

eBird data from 2006-2020. Estimated for 2020.

Fink, D., T. Auer, A. Johnston, M. Strimas-Mackey, O. Robinson, S. Ligocki, W. Hochachka, L. Jaromczyk, C. Wood, I. Davies, M. Iliff, L. Seitz. 2021. eBird Status and Trends, Data Version: 2020; Released: 2021. Cornell Lab of Ornithology, Ithaca, New York. https://doi.org/10.2173/ebirdst.2020

TheCornellLab   Data provided by **eBird**

Species Account   Red-headed Woodpeckers are beautiful, noisy, eye-catching birds. Perhaps for these reasons, they are often the "spark bird" for young bird-watchers, leading to a lifelong passion. The Red-headed Woodpecker has been given many colorful nicknames over the years, including the shirttail bird, the flag bird, and the flying checkerboard. The species was a symbol of war among Native Americans, with the blood-red heads of these woodpeckers used as battle ornaments by Plains tribes.

Red-headed Woodpeckers are unusual birds in several respects. They are one of only a handful of woodpecker species that cache their food, wedging acorns, seeds, and even large insects into nooks and crannies in trees to be eaten later. They are also more adept at catching insects on the wing than any other woodpecker in the eastern United States. In open forest or clearings they can be seen flying out from a tree, snatching a bug on the wing, and then returning to their perch. Perhaps this habit explains their preference for open forest or savanna-type habitat rather than heavily wooded tracts.

Red-headed Woodpeckers are omnivorous and have a more varied diet than other woodpeckers. The diet shifts seasonally but consists of approximately one-third animal material (largely insects such as cicadas, beetles, grasshoppers, and bees) and two-thirds plant material (acorns, beechnuts, and fruit). Red-headed Woodpeckers are occasionally seen hopping on the ground looking for food, and they may eat earthworms and, rarely, small rodents. In addition, they occasionally take other birds' eggs or nestlings.

Across the Southeast, Red-headed Woodpeckers are present year-round, the exception being in the higher elevations of the Appalachian Mountains. Red-headed Woodpeckers breed in the mountains but move to lower elevations for the winter. They also move somewhat nomadically east and west in some parts of the country. These movements may be in response to food availability, as mast (nut) production varies widely across the birds' range.

There are likely several reasons for the decline of the Red-headed Woodpecker, including collisions with vehicles and competition with other species for cavity nest sites. Wildfires once played a role in maintaining the open forest these birds need, as well as providing dead standing trees suitable for nesting. Selective thinning in woods and prescribed burning may improve the Red-headed Woodpecker's habitat by reducing the understory and creating dead snags for roosting and nesting.

Identification The Red-headed Woodpecker has a snow-white belly, black wings with white wing patches, and a bluish-gray beak and feet. In the Southeast, it is the only bird with a completely red head. The Red-bellied Woodpecker is similar in size but has red only on the back of its head. The Pileated Woodpecker has a red crest on its head but is much larger than the Red-headed Woodpecker. The plumage of male and female Red-headed Woodpeckers is identical. Juveniles are similar to adults but have brownish heads.

Vocalizations A sharp, raspy *cheer* is the most commonly heard call. Sometimes the birds give a lower, more rolling churr that is repeated rapidly. Red-headed Woodpeckers also drum on trees, branches, and occasionally houses.

Nesting Red-headed Woodpeckers can create a cavity suitable for nesting in about two weeks. The male does the majority of the excavation and usually chooses a dead or decaying tree or branch. The cavity can be more than a foot deep, with an entrance hole about two inches in diameter. Nest sites are often reused the following year. They can also be beneficial to other species of wildlife that use cavities but cannot create them. The female lays between three and ten pure white eggs that take about two weeks to hatch.

Red-headed Woodpeckers often perch and watch for flying insects. When a bird spots one, it flies out and snatches the insect on the wing before returning to its perch to eat.

Red-cockaded Woodpecker
(*Dryobates borealis*)

CONSERVATION CONCERN SCORE: 18 (High)

OTHER DESIGNATIONS: American Bird Conservancy watch list (Red),
2016 Partners in Flight Species of Continental Concern (Red), State Protected (AL),
State Special Concern (GA), State Endangered (MS, VA), Species of Greatest
Conservation Need (FL, KY, MD, NC, SC), Federally Endangered

ESTIMATED POPULATION TREND 1966–2019: −45%

SIZE: Length 9 inches; wingspan 14 inches

A Red-cockaded Woodpecker carefully inspects the
bark of a longleaf pine, looking for insects.

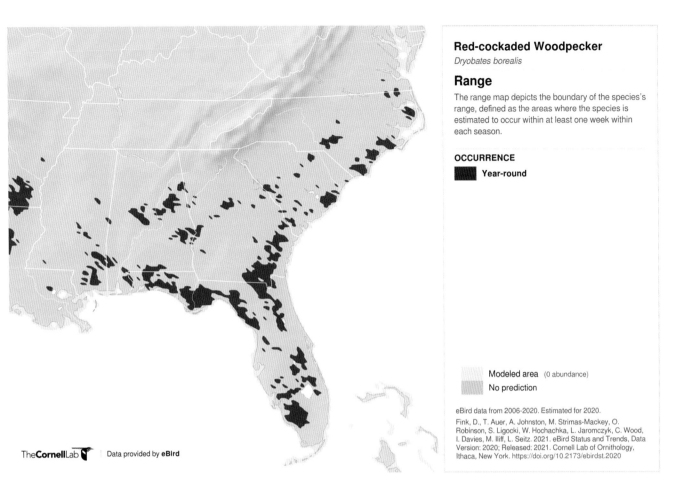

Red-cockaded Woodpecker

Dryobates borealis

Range

The range map depicts the boundary of the species's range, defined as the areas where the species is estimated to occur within at least one week within each season.

OCCURRENCE

■ Year-round

▨ Modeled area (0 abundance)
▨ No prediction

eBird data from 2006-2020. Estimated for 2020.

Fink, D., T. Auer, A. Johnston, M. Strimas-Mackey, O. Robinson, S. Ligocki, W. Hochachka, L. Jaromczyk, C. Wood, I. Davies, M. Iliff, L. Seitz. 2021. eBird Status and Trends, Data Version: 2020; Released: 2021. Cornell Lab of Ornithology, Ithaca, New York. https://doi.org/10.2173/ebirdst.2020

TheCornellLab | Data provided by eBird

Species Account The Red-cockaded Woodpecker inhabits the pine forests of the Southeast and is especially connected to the longleaf pine ecosystem that once covered vast swaths of that region of the United States. The male Red-cockaded Woodpecker has small red feathers (visible only when he is excited) at the edge of an extensive white facial patch. These red feathers were reminiscent of 1800s-era hat decorations called "cockades," giving the species its name.

Records from the 1800s indicate that Red-cockaded Woodpeckers once occurred as far north as Pennsylvania and New Jersey and, more recently, in Missouri, Kentucky, and Maryland. The species has now been extirpated from all these states. It is likely gone from Tennessee as well, and only very small populations hang on in Virginia. For nesting sites, the Red-cockaded Woodpecker needs mature (at least eighty to one hundred years old), living pine trees (but preferably suffering from heart rot, which makes it easier to excavate a cavity) and an open, fire-maintained understory. Modern forest practices have resulted in even-aged stands that are often harvested before they reach sufficient age to accommodate the woodpecker's needs, leading to significant population declines. Excavating a cavity can take anywhere from one to six years to complete. Although heart rot makes this process easier, the birds require living trees because they also drill a series of wells above and below the nest cavity. These wells fill and overflow with sap, leaving a sticky, resinous barrier that prevents tree-climbing snakes from reaching the nest, protecting the eggs and young from a significant predator.

Red-cockaded Woodpeckers often forage together in family groups, with the males found high on tree trunks or out on branches and the females lower on trunks. Up to four offspring from previous nesting seasons (usually males) stay with the adult birds and help raise additional young by sharing the incubating and feeding duties. The family group typically has a series of cavities in close proximity, called a cluster, where they roost, nest, and raise young. As the family group forages together during the day, they often peel bark off pines in search of insect prey, revealing a smoother, reddish surface. Trees with this smooth, reddish appearance indicate the birds' presence in the area. Ants make up a significant portion of the diet, sometimes up to 70 percent. Other prey items include insect larvae, termites, southern pine beetles, and bark beetles. Pine seeds and several types of fruit, including blueberries, poison ivy berries, and the fruit of the black cherry tree, are also consumed but typically constitute only a small portion of the diet. The birds rarely visit suet feeders.

Listed as endangered under the Endangered Species Conservation Act of 1969, the Red-cockaded Woodpecker has benefited from a host of costly management actions by the US Forest Service, the US Fish and Wildlife Service, and numerous state wildlife agencies and other conservation partners. Prescribed burning and the removal of hardwood vegetation have helped restore the savanna-like structure characteristic of longleaf pine forests prior to fire suppression and modern timber harvesting practices. Artificial nest cavities have been installed in many locations, and translocation of birds has helped maintain healthy genetics in declining populations. Despite this work, the birds' range continues to shrink, and the overall population is estimated at only seventy-eight hundred family clusters, with much of the remaining population in small, isolated groups susceptible to further decline. Even the strongest populations are at risk from storms, as evidenced by the devastating impact of Hurricane Hugo in September 1989. The storm hit South Carolina and decimated one of the last strongholds of the Red-cockaded Woodpecker: the Francis Marion National Forest population.

In late 2020 the US Fish and Wildlife Service proposed downgrading the Red-cockaded Woodpecker from endangered to threatened, arguing that, thanks to conservation efforts across its range, the species was no longer threatened with extinction. However, the agency's own recovery goals for downgrading have apparently not been met. It is difficult to balance the needs of the birds with the needs of local communities and the timber industry, not to mention the politics swirling around endangered species law. No doubt there will be much more debate on this topic.

Identification The Red-cockaded Woodpecker's back is black, barred or heavily speckled with white. The most distinguishing field mark is the large white cheek patch bounded by a black crown and a heavy black malar stripe. The sexes are identical except for the male's small red feathers at the top edge of the white cheek patch, which are rarely visible in the field.

Vocalizations Alexander Wilson (1831) compared the vocalizations of Red-cockaded Woodpeckers to "the chirping of young nestlings."

Nesting Three or four white eggs are laid, and the adult male incubates them during the night. Once they hatch, both parents feed the young, often with the help of offspring from previous nesting seasons.

This Red-cockaded Woodpecker (likely a male) has been color banded to help researchers monitor breeding success, habitat use, and overall survival rates.

American Kestrel
(*Falco sparverius*)

CONSERVATION CONCERN SCORE: 11 (Moderate)

OTHER DESIGNATIONS: 2021 USFWS Birds of Conservation Concern,
State Protected (AL), Special Concern (GA), State Threatened (FL),
Species of Greatest Conservation Need (KY, MD, MS, NC, PA, SC, WV)

ESTIMATED POPULATION TREND 1966–2019: −53%

SIZE: Length 9 inches; wingspan 22 inches

The adult male American Kestrel is beautifully colored,
with brick red and slaty blue on the back and wings.

American Kestrel *Falco sparverius*

Range

The range map depicts the boundary of the species's range, defined as the areas where the species is estimated to occur within at least one week within each season.

OCCURRENCE

- **Year-round**
- **Breeding season** May 17 - Aug 10
- **Non-breeding season** Nov 16 - Feb 22
- **Pre-breeding migratory season** Mar 1 - May 10
- **Post-breeding migratory season** Aug 17 - Nov 9

Note: Seasonal ranges overlap and are stacked in the order above; view full range in season maps.

SEASONS TIMELINE

J F M A M J J A S O N D

Modeled area (0 abundance)
No prediction

eBird data from 2006-2020. Estimated for 2020.

Fink, D., T. Auer, A. Johnston, M. Strimas-Mackey, O. Robinson, S. Ligocki, W. Hochachka, L. Jaromczyk, C. Wood, I. Davies, M. Iliff, L. Seitz. 2021. eBird Status and Trends, Data Version: 2020; Released: 2021. Cornell Lab of Ornithology, Ithaca, New York. https://doi.org/10.2173/ebirdst.2020

TheCornellLab | Data provided by **eBird**

Species Account The American Kestrel is the smallest and most widespread falcon in the United States. It is often seen perched on power lines in open country, or it may be spotted riding the air currents and flapping to hold itself in place as it searches the ground for prey. The distinctive *klee klee klee* call is often the first sign of the birds' presence in cattle pastures, savannas, parklands, airports, and other areas with short grass and few trees from Alaska to South America.

Kestrels are year-round residents in the Appalachians and the Southeast, although their winter numbers are inflated by birds that nest farther north in the United States and Canada and then migrate south. Males usually arrive on their wintering grounds about ten days later than females. In the fall, the passage of kestrels can be an impressive sight along the Atlantic coast or in the Appalachian passes that funnel migrating birds, such as Hawk Mountain in Pennsylvania. Kestrel numbers typically peak in late September at Hawk Mountain.

Once commonly known as the Sparrow Hawk, the American Kestrel is actually much more likely to take insects or small rodents than other birds, which make up

The adult female American Kestrel is solid rust with black
stripes on the back but shares the male's masked face pattern.

less than 10 percent of the typical diet. Grasshoppers, cicadas, dragonflies, and other
insects are important food items for kestrels, especially during the breeding season. In
winter, rodents make up a larger portion of the diet. Interestingly, it may be possible
for kestrels to identify locally high rodent populations based on their urine trails; these
trails reflect ultraviolet light, which is visible to the birds. Other prey items include
lizards, small snakes, bats, and fish. Kestrels usually consume insects on the spot, but
they may fly to a low perch to eat larger prey. In some cases, kestrels cache food in tree
cavities or clumps of grass and return to eat it later.

A lack of nest cavities is apparently the most important factor in the American Kestrel's overall population decline. Although these birds prefer open habitats, they need
some scattered trees large enough to contain nesting cavities. As agricultural practices
have shifted toward larger, "cleaner" farms with fewer hedgerows, populations of this
beautiful little falcon have declined. The nonnative European Starling also competes
with the American Kestrel for suitable nest sites. Kestrels will nest in old barns and other
outbuildings, and they accept nest boxes built to suit their needs. In some places, local
populations have greatly increased because of the efforts of a few dedicated individuals
who have installed and maintained a network of nest boxes specifically designed for
these birds (which can be purchased online).

This adult male American Kestrel was making a steep dive when captured in this photograph. Seconds later, he emerged from the edge of a grassy field with a mouse and flew to a nearby cornstalk to consume it.

Identification The male American Kestrel has a buffy breast, slate-blue wings and head, black facial markings, brick-red back and tail, and black spots on the belly, wings, and back. The tail has a thick black band near the tip. The female is rusty colored on the back and tail with black barring, a slate-blue head, and black facial markings; the belly is buffy with brown streaks. Both sexes have black spots on the back of the head and a rusty patch on the crown. The female is larger and heavier than the male.

Vocalizations The most common vocalization is a shrill *klee klee klee* given in flight or from a perch.

Nesting Nests are placed ten to thirty feet high. The female makes a shallow depression in whatever wood chips or other debris is available in the bottom of the cavity; no additional nesting material is used. Four to six pale, spotted eggs are laid. For the first ten days or so after hatching, the male brings food to the female and the nestlings. Then both parents hunt to feed the chicks.

Loggerhead Shrike
(*Lanius ludovicianus*)

CONSERVATION CONCERN SCORE: 11 (Moderate)

OTHER DESIGNATIONS: 2021 USFWS Birds of Conservation Concern,
In Need of Management (TN), State Special Concern (GA, NC),
State Threatened (VA), State Endangered (MD), Species of Greatest Conservation
Need (AL, FL, GA, KY, MS, PA, SC, WV)

ESTIMATED POPULATION TREND 1966–2019: –76%

SIZE: Length 9 inches; wingspan 12 inches

A Loggerhead Shrike hunting from an exposed perch. The bird's
hooked upper mandible is useful for damaging a victim's spine.

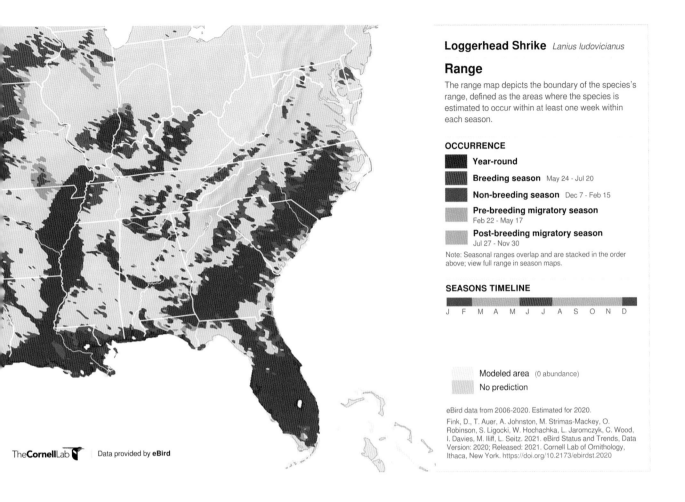

Loggerhead Shrike *Lanius ludovicianus*

Range

The range map depicts the boundary of the species's
range, defined as the areas where the species is
estimated to occur within at least one week within
each season.

OCCURRENCE

- **Year-round**
- **Breeding season** May 24 - Jul 20
- **Non-breeding season** Dec 7 - Feb 15
- **Pre-breeding migratory season**
  Feb 22 - May 17
- **Post-breeding migratory season**
  Jul 27 - Nov 30

Note: Seasonal ranges overlap and are stacked in the order
above; view full range in season maps.

SEASONS TIMELINE

J F M A M J J A S O N D

Modeled area (0 abundance)

No prediction

eBird data from 2006-2020. Estimated for 2020.
Fink, D., T. Auer, A. Johnston, M. Strimas-Mackey, O.
Robinson, S. Ligocki, W. Hochachka, L. Jaromczyk, C. Wood,
I. Davies, M. Iliff, L. Seitz. 2021. eBird Status and Trends, Data
Version: 2020; Released: 2021. Cornell Lab of Ornithology,
Ithaca, New York. https://doi.org/10.2173/ebirdst.2020

TheCornellLab | Data provided by eBird

Species Account Like a falcon trapped in a songbird's body, Loggerhead Shrikes are
amazingly fierce. Only about the size of a cardinal and possessing no talons, these birds
are still capable of killing and carrying prey equal to their own size and weight. They
prefer open habitat with widely scattered thorny trees intermixed with short-grass areas
or barren ground, and they often perch while scanning for prey. Shrikes consume a
wide variety of quarry, and usually the height of the perch correlates with the size of the
prey, with smaller animals being taken from lower perches. Insects typically dominate
the menu, and 60 to 80 percent of the diet consists of grasshoppers, crickets, beetles,
earwigs, ants, and other insects.

Although it is more common for shrikes to perch and watch for prey from above, the
birds sometimes stalk insects on the ground by half opening their wings, dragging the
tips of their primary feathers, and flashing the large white patches on their wings. It is
unclear what purpose this serves, although Northern Mockingbirds, which also have
white wing patches, use similar tactics. Some suggest that this behavior may startle the
insects, making them easier to locate and capture.

During the breeding season, while the adults are feeding the young, and during the winter when insects are less abundant, larger prey is taken, including birds and small mammals such as mice, shrews, and voles. Lizards, frogs, and small snakes also make up significant portions of the diet in some areas. When targeting these larger vertebrates, shrikes usually attack from behind and attempt to grab or strike the neck area. Projections on the bird's upper mandible called tomial teeth may help break the neck vertebrae or damage the spinal cord, leaving the animals paralyzed and defenseless. The shrike then carries the immobilized animal to a thorny tree or a barbed wire fence, where it impales its victim to make it easier to tear apart and consume. A favored tree or fence line with a cache of impaled prey items is referred to as a pantry or larder. These food storage areas may be helpful in attracting a mate and can be important when food becomes scarce. This behavior of hanging food on sharp objects or wedged into the crooks of branches has earned the shrike the nicknames "Thornbird" and "Butcherbird." It has also led to persecution by humans, including shooting the birds, despite the fact that the insects and rodents they eat are considered pests in many areas.

Loggerhead Shrikes are still fairly common across the Gulf coast states, portions of the lower Mississippi River floodplain, and north through Georgia and the Carolinas. However, they are now almost completely absent from the Northeast and are present in rapidly dwindling numbers in the Midwest. Of the world's thirty species of shrikes, the Loggerhead Shrike is the only one found exclusively in North America, where it has declined by more than 75 percent since the 1960s. More intensive agriculture leading to a reduction in hayfields, pastures, and old field habitat and the loss of hedgerows between fields may partially explain the decline. The widespread use of insecticides and other agricultural chemicals has likely played a role as well. Additional causes include habitat loss due to surface coal strip mining, urban development, and the invasion of nonnative fire ants and the chemicals used to control them.

Identification Loggerhead Shrikes are large-headed birds with a heavy, hooked black bill, a black face mask, and a gray body. The wings are black with large white patches that are especially visible in flight. When perched, they appear slim bodied and often hold their tails horizontal to the ground. Males are slightly bigger than females, but otherwise, the sexes are similar. Loggerhead Shrikes are slightly smaller than the very similar Northern Shrike and have a thicker black mask.

Vocalizations Calls consist of a series of harsh clicks, rattles, and other rasping notes that sound almost scolding in nature.

Nesting Nests are constructed by the female and are usually fairly close to the ground (less than ten feet above it). In Alabama, nest initiation peaks in early April. Favorite locations include hawthorn trees, junipers, and multiflora rose bushes. Nests are woven together, and the cup is well insulated with feathers, moss, fur, or other material. Five or six light-brown, heavily speckled eggs are laid. In some years, second and third broods are attempted.

This Loggerhead Shrike is consuming a cricket it spotted while perched on a power line and then swooped down and captured.

Florida Scrub-Jay
(*Aphelocoma coerulescens*)

CONSERVATION CONCERN SCORE: 20 (High)

OTHER DESIGNATIONS: American Bird Conservancy watch list (Red), 2021 Audubon Priority Birds list, 2016 Partners in Flight Species of Continental Concern (Red), Species of Greatest Conservation Need (FL), Federally Threatened

ESTIMATED POPULATION TREND 1966–2019: Not available

SIZE: Length 11 inches; wingspan 14 inches

The Florida Scrub-Jay is crestless, with a triangular gray patch on the back and a bright blue head, wings, and tail.

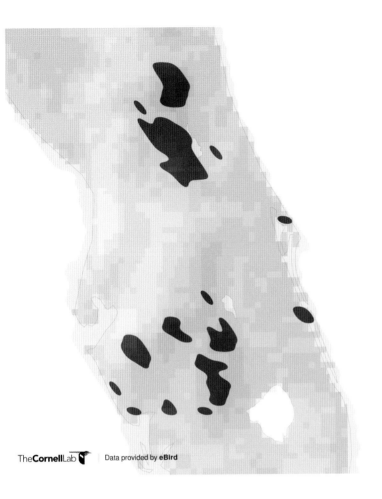

Florida Scrub-Jay

Aphelocoma coerulescens

Range

The range map depicts the boundary of the species's range, defined as the areas where the species is estimated to occur within at least one week within each season.

OCCURRENCE

▓ Year-round

☐ Modeled area (0 abundance)
▒ No prediction

eBird data from 2006-2020. Estimated for 2020.

Fink, D., T. Auer, A. Johnston, M. Strimas-Mackey, O. Robinson, S. Ligocki, W. Hochachka, L. Jaromczyk, C. Wood, I. Davies, M. Iliff, L. Seitz. 2021. eBird Status and Trends, Data Version: 2020; Released: 2021. Cornell Lab of Ornithology, Ithaca, New York. https://doi.org/10.2173/ebirdst.2020

Species Account Found nowhere else on the globe, the Florida Scrub-Jay is at home in the head-high oak- and palmetto-dominated scrublands of central Florida. It appears to labor in flight and seems highly averse to straying outside of its preferred habitat. In fact, it is not unusual for these birds to spend their entire lives within a mile of the shrub that held the nest they were born in. Rather than dispersing to establish their own territories, young birds often stay with their parents and up to five other siblings. This family group works together to defend its territory and feed nestlings. One member of the group is always on the lookout for predators and alerts the others with a specific call, depending on whether the threat comes from a snake (a common cause of egg loss and mortality of young birds) or some other source. This type of cooperative breeding structure allows a pair to successfully raise more young than would be possible on their own.

Scrub-jays depend on fires to maintain their preferred habitat, as burning creates openings in the vegetation, leaving areas of bare sand. The birds use these open sandy spots to bury acorns harvested in the fall. A single jay can hide between sixty-five hundred and eight thousand acorns in a season. The birds return periodically to dig up and feed on the acorns, especially when food is scarce, or just to check on their location and condition. Fires caused by lightning strikes used to be a common occurrence in the scrub habitats of central Florida, where the sandy soils support the quick drying of fuel to feed the fires. In addition to scrub-jays, many of the native plants rely on fire, which they are able to survive thanks to thick, insulating bark and deep roots. Following a blaze, the vegetation quickly responds with lush growth and greater productivity. However, if an area goes too long without a fire, the habitat shifts and becomes overgrown with thick vegetation, providing places for snakes and hawks to hide. After about twenty years without a fire, the habitat is no longer suitable for Florida Scrub-Jays.

Appropriate habitat for Florida Scrub-Jays is very patchy in distribution. There are many small satellite populations across the state's central peninsula, with a handful of larger, more stable populations where the habitat is managed, such as the Archbold Field Station and Merritt Island National Wildlife Refuge. Despite their small size, these isolated populations are quite important because the birds there provide genetic diversity, which is vital even for the large, well-managed populations. Although it is extremely rare for birds to travel more than eight miles from their place of birth, some gene flow occurs between populations. With such a small number of birds remaining and a very localized breeding strategy, inbreeding is a problem, so it is important to have some mixing of genes from birds in nearby populations. Excessive inbreeding results in lower hatching rates and lower overall survival rates of young birds, even if all other conditions are good.

Today, the total population of Florida Scrub-Jays is estimated at less than five thousand birds. In large part, this is due to habitat loss, with only about 10 percent of the original scrub habitat remaining in central Florida. Over time, this habitat has been lost to residential development as Florida's human population has boomed. Agricultural conversion of scrub habitat to citrus groves and other uses has also reduced and fragmented the best habitats. The few remaining areas of high-quality habitat have been impacted by the lack of naturally occurring fires, as many lightning strikes occur where there is no longer any natural fuel left to burn or the fire is immediately extinguished to protect residential areas or other infrastructure. Overall, this leaves the scrub-jay in a precarious position, with low population levels, little genetic diversity, and high susceptibility to hurricanes or other adverse events. It is estimated that the remaining population dropped by 25 percent between 1983 and 1993, and it has likely continued to decline since then.

Identification The Florida Scrub-Jay is a large, long-tailed, flat-headed bird with powerful legs and feet and a dark, heavy, chisel-like bill. It has a bright blue head, tail, and wings with a triangular gray patch on its back. The whitish throat is bordered by a partial bib, and the birds sport a white or gray supercilium and belly. Unlike the Blue

Jay, the Florida Scrub-Jay does not have a crest. It is similar in appearance to other scrub-jays that live in the western United States but is distinctive in the limited area where it occurs. It can be quite tame and frequently allows a close approach.

Vocalizations Florida Scrub-Jays are capable of a wide array of vocalizations. Some calls, such as the *hiccup* call, are given only by the female. Most calls are harsh or rattling. Vocalizations vary across populations, even those that are separated by twenty miles or less.

Nesting Sand live oak is the preferred location for the nest, which is usually built only a few feet off the ground and is often placed under a clump of heavier vegetation such as a vine. The nest is built by both sexes, but the female is largely responsible for shaping the cup. The eggs are greenish with reddish-brown speckles that are more numerous on the larger end. Clutches can consist of one to six eggs but typically contain three or four. Second clutches are laid if the first nest fails.

Florida Scrub-Jays are capable of caching thousands of acorns in sandy openings within the scrub forest. They return to these caches throughout the year, either to consume the acorns or simply to check on them.

Brown-headed Nuthatch
(*Sitta pusilla*)

CONSERVATION CONCERN SCORE: 13 (Moderate)

OTHER DESIGNATIONS: 2021 USFWS Birds of Conservation Concern,
   2021 Audubon Priority Birds list, Species of Greatest Conservation Need (FL, MS,
   NC, SC, TN)

ESTIMATED POPULATION TREND 1966–2019: −15%

SIZE: Length 5 inches; wingspan 8 inches

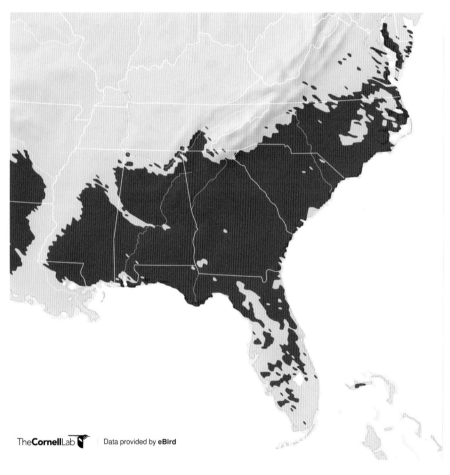

Brown-headed Nuthatch *Sitta pusilla*

Range

The range map depicts the boundary of the species's range, defined as the areas where the species is estimated to occur within at least one week within each season.

OCCURRENCE

■ Year-round

▨ Modeled area (0 abundance)
▥ No prediction

eBird data from 2006-2020. Estimated for 2020.

Fink, D., T. Auer, A. Johnston, M. Strimas-Mackey, O. Robinson, S. Ligocki, W. Hochachka, L. Jaromczyk, C. Wood, I. Davies, M. Iliff, L. Seitz. 2021. eBird Status and Trends, Data Version: 2020; Released: 2021. Cornell Lab of Ornithology, Ithaca, New York. https://doi.org/10.2173/ebirdst.2020

TheCornellLab | Data provided by **eBird**

Species Account The squeaky call of the Brown-headed Nuthatch has been likened to the sound of a dog's chew toy or a rubber ducky being squeezed. Because it is so distinctive and easily recognized, listening for their call may be the best way to find these tiny birds as they scramble along limbs or climb up and down tree trunks peering into and probing nooks and crevices for food. This species is truly a denizen of the longleaf, loblolly, shortleaf, and slash pine forests of the southeastern United States, and nearly the entire population is contained within US borders—with the exception of a dwindling population in the Bahamas that may, in fact, be a subspecies.

During the warm months, the Brown-headed Nuthatch's diet consists primarily of insects—spiders, beetles, cockroaches, and others, as well as larvae and egg cases—which it finds under loose bark and other hiding spots. In the winter, food selection shifts to plant matter, with pine seeds making up 80 to 95 percent of the diet. Interestingly, the Brown-headed Nuthatch is one of the few bird species that uses tools to search for prey. They sometimes carry a small piece of bark in their beaks and use it to pry up the loose edges of peeling bark so they can look for insects hiding underneath. Nuthatches have also been observed using bark flakes to cover up caches of stored pine seeds, likely in an effort to prevent others from stealing them.

The Brown-headed Nuthatch is apparently nonmigratory, and pairs may use the same territory through multiple nesting seasons. In some cases, offspring from the previous nesting season stay with the adults and help defend the territory and feed nestlings. The best habitat for these birds appears to be mature pine forest with low levels of woody vegetation in the understory. Living pines are needed for foraging, but dead snags are also an important component of the habitat, as that is where the birds usually choose to excavate their nest cavities. This type of habitat is typically created by frequent low- to moderate-intensity fires, although it can be simulated by certain types of timber harvesting practices or even in residential subdivisions, golf courses, or parks with a scattering of large pine trees, few shrubs in the understory, and some dead trees for nest cavity construction if there are no birdhouses or previously excavated cavities available.

Much of the decline in the Brown-headed Nuthatch population is likely due to the loss of habitat. Large-scale clear-cutting of forests, management practices that harvest younger, same-aged trees, conversion of pine forests to mixed forest types, habitat fragmentation, and fire suppression are all likely to result in further declines of this species. Because populations are nonmigratory and are tied long term to specific territories, it may be necessary to translocate birds to reestablish populations where they have disappeared, since they are unlikely to recolonize suitable habitat on their own. Such a translocation project has been attempted in parts of Florida (where this species has declined approximately 54 percent since 1966), with some success.

Identification The Brown-headed Nuthatch is a small, active bird that can be seen clinging to the bark of tree trunks and branches in pine forests of the Southeast. Unlike woodpeckers, the nuthatch does not use its tail for support but relies on its strong feet to maintain its grip—even when foraging upside down. The birds have a blue-gray back and flanks, a creamy throat, a brown head, and a small white spot on the nape of the neck.

Vocalizations Brown-headed Nuthatches have high-pitched, squeaky calls that are unlike those of any other birds in their range. Their calls are easily distinguished and very helpful in locating these birds, which spend a good part of their time high up in mature trees.

Nesting Nests are often placed fairly low to the ground—occasionally as low as five feet. Nests are constructed of grass, bark, Spanish moss, and other plant material, as well as hair, cocoon fibers, and feathers. The female lays four to six white or cream-colored eggs with reddish-brown speckles. The pair may excavate multiple nest cavities together before settling on one.

Tiny Brown-headed Nuthatches are in constant motion as they clamber up and down tree trunks or out onto pine branches in search of food.

Wood Thrush
(*Hylocichla mustelina*)

CONSERVATION CONCERN SCORE: 14 (High)

OTHER DESIGNATIONS: American Bird Conservancy watch list (Yellow),
 2021 USFWS Birds of Conservation Concern, 2016 Partners in Flight Species
 of Continental Concern (Yellow), In Need of Management (TN),
 Species of Greatest Conservation Need (FL, KY, MD, MS, PA, SC, VA, WV)

ESTIMATED POPULATION TREND 1966–2019: −51%

SIZE: Length 8 inches; wingspan 13 inches

The Wood Thrush is one of the most beautiful singers in
the bird world. Its lovely chorded notes light up the dark,
shady woodlands of the Appalachians and the Southeast.

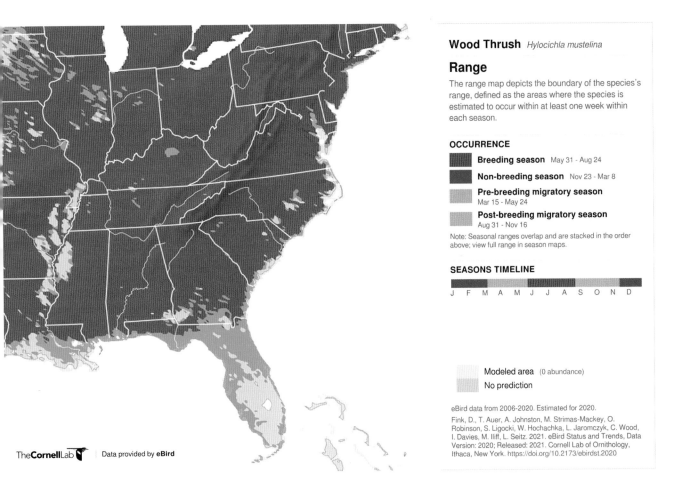

Wood Thrush *Hylocichla mustelina*

Range

The range map depicts the boundary of the species's range, defined as the areas where the species is estimated to occur within at least one week within each season.

OCCURRENCE

Breeding season May 31 - Aug 24

Non-breeding season Nov 23 - Mar 8

Pre-breeding migratory season
Mar 15 - May 24

Post-breeding migratory season
Aug 31 - Nov 16

Note: Seasonal ranges overlap and are stacked in the order above; view full range in season maps.

SEASONS TIMELINE

J F M A M J J A S O N D

Modeled area (0 abundance)

No prediction

eBird data from 2006-2020. Estimated for 2020.

Fink, D., T. Auer, A. Johnston, M. Strimas-Mackey, O. Robinson, S. Ligocki, W. Hochachka, L. Jaromczyk, C. Wood, I. Davies, M. Iliff, L. Seitz. 2021. eBird Status and Trends, Data Version: 2020; Released: 2021. Cornell Lab of Ornithology, Ithaca, New York. https://doi.org/10.2173/ebirdst.2020

TheCornellLab | Data provided by **eBird**

Species Account The enchanting song of the Wood Thrush can be heard drifting out of shaded woodlands in the Southeast and across much of the eastern United States. Because of a dual-chambered voice box, the Wood Thrush can make its own harmonies by singing two different notes at the same time. The result is a beautiful but melancholy, chorded, flute-like song. In an 1853 entry in his personal journal, Henry David Thoreau wrote of the Wood Thrush, "This is the only bird whose note affects me like music. It lifts and exhilarates me. It is inspiring. It changes all hours to an eternal morning."

The song of the Wood Thrush has three separate parts. The first phase (A) consists of several short introductory notes, the middle phase (B) is the familiar *ee-oo-lay,* and the final phase (C) consists of pairs of notes sung at the same time, creating a chord. Male Wood Thrushes may have one to three versions of part A, two to eight versions of part B, and six to twelve versions of part C. Each of these versions is interchangeable, giving the Wood Thrush a wide variety of possible combinations. Some males may sing more than fifty variations, cycling through many different songs during their dawn chorus. Interestingly, male Wood Thrushes removed from the wild at birth are able to

Like many thrushes, the Wood Thrush spends a fair bit of time on the ground, often searching among the leaf litter for insects or other food.

vocalize the A and C portions of the song but sing only short, slurred versions of the B phase, indicating that they may learn this part by listening to other Wood Thrushes.

Ideal Wood Thrush habitat includes large blocks (more than five hundred acres) of relatively unfragmented, mature forest with trees taller than fifty feet, a moderate understory of shrubs and saplings, and a relatively open forest floor with water nearby. Wood Thrushes often feed on the forest floor, where they flip leaves over in search of prey. Common food items include beetles, spiders, caterpillars, and ants, although they occasionally take larger prey such as small salamanders. In late summer, the birds eat a variety of fruits, including spicebush, blueberry, holly, grapes, jack-in-the-pulpit, Virginia creeper, and dogwood.

Although the Wood Thrush sometimes attempts to nest in smaller, fragmented woodlots, these habitats are often "sinks," where the birds may nest but typically cannot produce as many young as in higher-quality habitats. Increased parasitism from the Brown-headed Cowbird in fragmented woodlands may explain the Wood Thrush's lower reproductive success. One study published in the *Wilson Bulletin* reported a cowbird parasitism rate of 42.1 percent (Hoover and Brittingham 1993). However, other sources indicate that in some forests with high fragmentation, nearly every Wood Thrush nest is parasitized by cowbirds. Despite these impacts, the role of forest fragmentation on the Wood Thrush is complex. In some cases, Wood Thrush nestlings seem to grow faster when the nest is placed near some type of disturbance (e.g., clear-cuts, power lines) within a larger forest block. It is thought that the faster growth rate is due to higher food production in these openings than in the surrounding mature forest.

Identification The Wood Thrush has a body shape similar to that of the American Robin, but it is slightly smaller. The back is a rich reddish brown, with brighter red hues at the nape. The belly is white with large dark spots. Most birds show at least a partial white eye ring. The similar Hermit Thrush has the rich red color only on its tail, while the back and wings are light brown. The Hermit Thrush's spots are also smaller and less numerous than those of the Wood Thrush. The larger Brown Thrasher has a longer tail than the Wood Thrush and has streaks instead of spots on the belly.

Vocalizations The song is a beautiful, flute-like *ee-oo-lay*, usually given while perched on a bare branch. The call is a sharp *pit-pit-pit*, which can be quite loud and forceful.

Nesting Nest height varies, but it is often placed in a sapling less than ten feet off the ground. Common species used for nesting include flowering dogwood, spicebush, and several oak species. The female lays three or four pale greenish-blue eggs. The young fledge within twelve days after hatching and are tended by both parents. Each pair usually attempts multiple nests per year.

Bachman's Sparrow
(*Peucaea aestivalis*)

CONSERVATION CONCERN SCORE: 16 (High)

OTHER DESIGNATIONS: American Bird Conservancy watch list (Red),
2021 USFWS Birds of Conservation Concern, 2016 Partners in Flight Species of
Continental Concern (Red), State Protected (AL), State Special Concern (GA, NC),
State Threatened (VA), State Endangered (TN), Species of Greatest Conservation
Need (FL, KY, SC)

ESTIMATED POPULATION TREND 1966–2019: −73%

SIZE: Length 6 inches; wingspan 7 inches

Palmetto is common in the understory of Bachman's Sparrow habitat,
along with a variety of grasses and sedges under a canopy of pine.

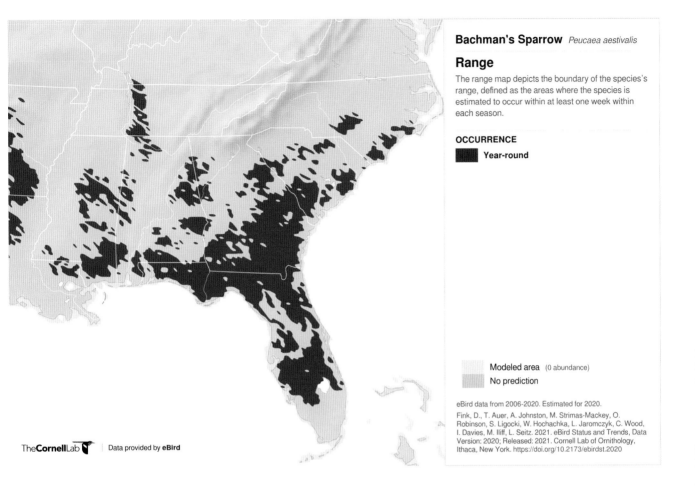

Species Account The Bachman's Sparrow was first described by John James Audubon in 1834, from a specimen he collected in South Carolina. Audubon named the species in honor of John Bachman, a clergyman from Charleston who hosted Audubon during one of his American expeditions. It is a rather nondescript sparrow characteristic of the pine woods of the southeastern United States, and its Latin name makes reference to its preferred habitat: loosely translated, it means "pine woods finch of the summer." Even in areas where the birds are resident year-round, they are extremely difficult to find in the nonbreeding season, when they often stay hidden on the ground or in clumps of grass. It is only during the breeding season, when the males sing to defend and establish their territories, that the bird is more easily located.

This species prefers to stay on the ground to forage for insects and seeds. According to a study done in Alabama, its diet consists of 58 percent insects and 42 percent plant material (Weston 1968). Favorite prey items include weevils, caterpillars, and beetles. Seeds from a variety of sedges and grass species make up the remainder of the diet, with grasses in the genus *Panicum* being especially important.

In the late 1800s and early 1900s, the Bachman's Sparrow experienced a significant range expansion to the north, likely in response to extensive timber harvesting that transformed numerous areas to grassy, early successional habitat. By 1920, it was considered a breeding bird in Pennsylvania, West Virginia, Ohio, Indiana, and Illinois. Beginning in the 1930s, however, its range began to contract as these newly cleared habitats were converted to agriculture, developed as residential areas, or allowed to grow back into dense forest. This declining population trend continues to the present day. The Bachman's Sparrow is now nearly or completely extirpated from the five states mentioned above, as well as Maryland and Missouri. Virginia, Kentucky, and Tennessee have only small, isolated populations at best.

The core of the birds' present range from the Carolinas through Georgia and the Gulf coast states has seen significant population declines in recent decades, largely due to the suppression of naturally occurring fires and logging practices that influence the age and structure of the remaining pine woods. This has resulted in a great reduction in the availability of the Bachman's Sparrow's preferred habitat, which consists of open, mature pine forest with a grassy understory that is largely devoid of a heavy shrub layer. Fire or some other form of disturbance is needed on a regular basis (typically, every three to five years) to keep the habitat in prime condition for the Bachman's Sparrow, except in areas where poor soil slows the growth of brushy species. Even where suitable habitat remains, however, the sparrows are often absent. This seems to indicate that the reasons for the birds' decline may be more nuanced than previously thought.

Identification The Bachman's Sparrow is a fairly long-tailed bird with a large head and beak. It is somewhat similar to a Field Sparrow in overall coloration and appearance. The birds are quite plain overall, with an off-white belly with no markings and a rufous cap. The back is grayish with rusty-brown markings. A reddish line extends behind the eye, and there is a faint malar stripe at the side of the throat. Males and females are not distinguishable in the field.

Vocalizations Hearing a singing male is the best way to locate this secretive species. The song starts with an introductory note followed by a trill at a different pitch. This can be confused with the song of the Rufous-sided Towhee, which inhabits similar habitats in some areas.

Nesting Nests are built on the ground, often beside or under a pine seedling, a clump of grass, or a palmetto leaf. Nests often have overhanging vegetation that helps with concealment. A typical clutch consists of four white, unmarked eggs. Across their current range, most females attempt two clutches.

Seen from the back, Bachman's Sparrows are a blend of rust and gray, which provides camouflage and makes it extremely difficult to find them if they are not singing.

Florida Grasshopper Sparrow
(*Ammodramus savannarum floridanus*)

CONSERVATION CONCERN SCORE: Not available. Although the Grasshopper Sparrow as a species scores a 12 (Moderate), the Florida Grasshopper Sparrow would score much higher.

OTHER DESIGNATIONS: Species of Greatest Conservation Need (FL), Federally Endangered

ESTIMATED POPULATION TREND 1966–2019: Not available

SIZE: Length 5 inches; wingspan 8 inches

Because the range of the Florida Grasshopper Sparrow is so small, it is not feasible to include a map. Generally, this subspecies is present in just a few counties in central Florida.

Species Account The Grasshopper Sparrow has a wide geographic range across the continent, stretching from the East to the Great Plains. It is absent from much of the Rocky Mountains and the arid Southwest but can be found breeding along a good portion of the California coast. It is generally accepted that four subspecies breed in North America, and although many populations appear to be declining due to loss of habitat, the Florida Grasshopper Sparrow seems to be in the most trouble.

First described in 1902, Florida Grasshopper Sparrows are found on large, treeless, frequently burned dry prairies across portions of south-central Florida. Common plant species in this habitat include bluestem grasses, Saint-John's-wort, wire grass, dwarf oak, and saw palmetto. Florida Grasshopper Sparrows do not tolerate trees that grow more densely than one per acre. Because the Florida Grasshopper Sparrow's range falls within the area that has the highest number of thunderstorm days in the continental United States, it is thought that frequent wildfires caused by lightning strikes helped maintain the treeless prairie habitat with generous amounts of bare ground that these birds require.

Rarely found outside of a few counties located north and west of Lake Okeechobee, this subspecies has experienced rapidly declining populations largely due to conversion of land for agriculture, lack of fire to maintain habitat quality and keep woody species in check, and the introduction of "improved forage" pasture grasses for livestock. Overall, more than 85 percent of the dry prairie habitat in Florida has been lost. A federally endangered species since its original listing in July 1986, the Florida Grasshopper Sparrow has been closely monitored for several decades and has been considered endangered by the state of Florida since 1977. Limited to just three known locations on public land and a handful of populations on privately owned property, the overall population declined rapidly between 1999 and 2012, despite management efforts. During this time, spring counts of singing males on the three publicly owned properties dropped from more than three hundred to seventy-five, with only one singing male recorded at one of the sites, making this one of the rarest birds in all of North America.

In 2013 a number of federal, state, and other conservation partners began a captive breeding program as part of the recovery effort for this species. In the wild, Florida Grasshopper Sparrows have a nesting success rate of only 11 percent. Although captive rearing can reduce the threats to young birds from disease, fire ants, and other predators, it is difficult to determine how well captive-reared birds will do once they are released to the wild. The first 106 captive-reared birds were released in 2019 at the site with the largest remaining wild population, and a good number of them survived to the next spring and were observed holding territories and apparently attempting to breed. An additional fifty birds were released in 2020. If the population at this site can be stabilized, it will be important to supplement the population at the other sites as well, to avoid having so many of the remaining birds tied to just one location, which makes them susceptible to disease, storms, or other events that could be devastating to their chances for survival.

Identification   The Florida Grasshopper Sparrow is a small, short-tailed, flat-headed sparrow of the dry prairies. It is very similar overall to the Grasshopper Sparrow found throughout much of the East, but the Florida subspecies is smaller, with a heavier bill. It is also darker and has less reddish coloration in the feathers across the back; the edges of these dark feathers are more whitish than buffy.

Vocalizations   The song of the Florida Grasshopper Sparrow has been described as one of the weakest of any North American bird. A short series of three introductory notes is followed by an insect-like buzz or trill.

Nesting   Florida Grasshopper Sparrows nest on the ground, often under palmettos or near grass clumps. The nest is placed in a shallow scrape in the sandy substrate. The dome-shaped nests are made from narrow-leaved grasses. Three to five eggs make up the typical clutch, and the nestlings depart the nest about a week and a half after hatching.

The Florida Grasshopper Sparrow spends much of its time on the ground in open grasslands in central Florida. The treeless habitat is maintained by frequent fires.

Seaside Sparrow
(*Ammospiza maritima*)

CONSERVATION CONCERN SCORE: 14 (High)

OTHER DESIGNATIONS: American Bird Conservancy watch list (Yellow),
2021 USFWS Birds of Conservation Concern, 2016 Partners in Flight Species of
Continental Concern (Red), 2021 Audubon Priority Birds list, State Special Concern
(GA, MD), Species of Greatest Conservation Need (AL, MS, NC, VA)

ESTIMATED POPULATION TREND 1966–2019: +42%

SIZE: Length 6 inches; wingspan 8 inches

This Seaside Sparrow is *Ammospiza maritima
maritimus,* the gray, cold-toned subspecies found on
the northern Atlantic coast (New Jersey).

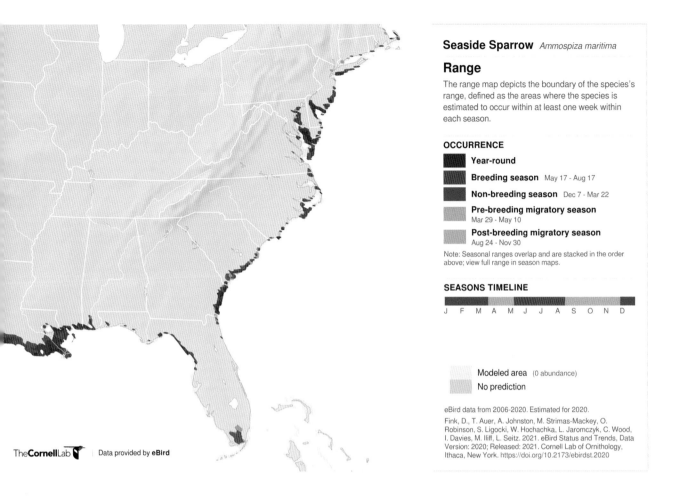

Seaside Sparrow *Ammospiza maritima*

Range

The range map depicts the boundary of the species's range, defined as the areas where the species is estimated to occur within at least one week within each season.

OCCURRENCE

Year-round

Breeding season   May 17 - Aug 17

Non-breeding season   Dec 7 - Mar 22

Pre-breeding migratory season   Mar 29 - May 10

Post-breeding migratory season   Aug 24 - Nov 30

Note: Seasonal ranges overlap and are stacked in the order above; view full range in season maps.

SEASONS TIMELINE

J F M A M J J A S O N D

Modeled area   (0 abundance)

No prediction

eBird data from 2006-2020. Estimated for 2020.

Fink, D., T. Auer, A. Johnston, M. Strimas-Mackey, O. Robinson, S. Ligocki, W. Hochachka, L. Jaromczyk, C. Wood, I. Davies, M. Iliff, L. Seitz. 2021. eBird Status and Trends, Data Version: 2020; Released: 2021. Cornell Lab of Ornithology, Ithaca, New York. https://doi.org/10.2173/ebirdst.2020

TheCornellLab | Data provided by eBird

Species Account The Seaside Sparrow is a grayish bird of the salt marshes of the Atlantic and Gulf coasts. Its proper taxonomy has been a topic of debate over the years, as populations have been considered three separate species—Seaside Sparrow, Dusky Seaside Sparrow, and Cape Sable Seaside Sparrow—or many different subspecies. Current thinking is that the Seaside Sparrow is one species with nine different subspecies (geographically distinct subpopulations that differ genetically or in physically observable ways) residing in salt or brackish marshes from southern Maine to the northern Gulf coast of Mexico. According to a 2005 article by Dr. James Rising, these subspecies are *Ammodramus* (now *Ammospiza*) *maritima maritimus,* the Northern Seaside Sparrow, from southern Maine and New Hampshire south to southern Virginia; *A. m. macgillivraii,* MacGillivray's Seaside Sparrow, from the coastal Carolinas and Georgia; *A. m. pelonota,* the extinct Smyrna Seaside Sparrow, from northeastern Florida; *A. m. nigrescens,* the extinct Dusky Seaside Sparrow, formerly from eastern Orange and northern Brevard Counties in Florida; *A. m. mirabilis,* the threatened Cape Sable Seaside Sparrow, which apparently no longer occurs in the Cape Sable region

The bird in this photograph, from an extensive coastal salt marsh in Georgia, is likely the *Ammospiza maritima macgillivraii* subspecies of Seaside Sparrow.

The Seaside Sparrow subspecies (*Ammospiza maritima fisheri*) from the western Gulf of Mexico (Texas Gulf coast) has warm buffy tones, especially on the breast.

of southwestern Florida but remains in certain freshwater marshes in the Everglades; *A. m. peninsulae,* Scott's Seaside Sparrow, of the Florida Gulf coast from Tampa Bay to Pepperfish Keys; *A. m. junicola,* the Wakulla Seaside Sparrow, in the Florida Panhandle from southern Taylor County to Escambia Bay; *A. m. fisheri,* the Louisiana Seaside Sparrow, from Pensacola, Florida, to San Antonio Bay, Texas; and *A. m. sennetti,* the Texas Seaside Sparrow, from Nueces and Copano Bays, Texas. The Atlantic and Gulf coast populations differ both genetically and in physical appearance. The plumage of Gulf coast birds tends to have a warmer, buffy wash compared with the cold gray colors of the Atlantic coast birds.

Because of their apparent sensitivity to disturbance, Seaside Sparrows may be a good indicator species for determining habitat quality in coastal salt marsh systems. For example, in marshes that have been ditched or where the natural hydrology has otherwise been altered, Seaside Sparrows require territories that are seven times larger than those of birds nesting in unaltered marshes. The birds are also sensitive to fire. Although fire is an important component of healthy salt marsh habitat because it keeps woody vegetation in check, Seaside Sparrows are typically absent from the burned area for a year after the fire. However, in the second year following a burn, the number of birds in the area may be triple the pre-fire population.

Although Breeding Bird Survey data appear to reflect a healthy increase in overall numbers since the 1960s, this species is not picked up on many survey routes, so the confidence in this population trend is not high. In fact, contrary to BBS data, during this same period, a number of local Seaside Sparrow populations have become extinct. This includes all the breeding populations from Florida's Atlantic coast and the Dusky Seaside Sparrow, which was found in the Merritt Island area of central Florida.

In 1973 a decision was made to classify the Dusky Seaside Sparrow as a subspecies of the Seaside Sparrow rather than continuing to consider it a unique species. This decision may have made it more difficult to attract the resources and attention necessary to protect the birds' remaining habitat. Following applications of the pesticide DDT and ditching operations in the area's marshes in an effort to control mosquitoes, along with the construction of new roads that altered the hydrology in these marshes, the Dusky Seaside Sparrow declined drastically. By the early 1980s, there were only six known birds (all males) remaining in the wild. Five of the six birds were captured in an effort to crossbreed the Dusky with Seaside Sparrow populations from Florida's Gulf coast. The last known Dusky Seaside Sparrow died in its cage in June 1987.

A similar story may be playing out with the Cape Sable Seaside Sparrow (*A. m. mirabilis*). Occurring only in south Florida, the Cape Sable Seaside Sparrow was apparently wiped out along the state's southwestern coast by a 1935 hurricane, but a very small population survives today in freshwater marshes primarily in Everglades National Park. Its existence in freshwater rather than saltwater marshes and its lighter greenish coloring make this subspecies unique among the other populations of Seaside Sparrow.

Preventing further upstream hydrologic alterations and managing the remaining habitat through the control of invasive species and appropriately timed and planned burns will be important factors in determining whether this population survives.

Identification Atlantic coast populations are generally gray with some indistinct streaking on the breast and back. Gulf coast birds have warmer buffy coloring, especially on the breast and face. All populations have relatively long bills, short tails, a clear white throat patch, and a yellow dot between the bill and the eye.

Vocalizations The song is buzzy and insect-like. It consists of a series of clicks and buzzes that are too complex for the human ear to fully differentiate. There are as many as eight different call notes given by both sexes. These calls are often used in territorial disputes or other aggressive interactions.

Nesting The nest is typically located at the ground level or within a few feet of the ground. It is constructed by the female alone and consists of a woven cup of grasses. Favored habitats include spartina, rushes, and salt grass. The nest is often covered with a loose canopy of vegetation. Three or four whitish, heavily speckled eggs are laid. Females have been known to overlap broods by initiating a second nest before the first clutch has fully fledged.

Saltmarsh Sparrow
(*Ammospiza caudacuta*)

CONSERVATION CONCERN SCORE: 19 (High)

OTHER DESIGNATIONS: American Bird Conservancy watch list (Red),
2021 USFWS Birds of Conservation Concern, 2016 Partners in Flight Species of
Continental Concern (Red), 2021 Audubon Priority Birds list, State Special Concern
(GA, MD), Species of Greatest Conservation Need (FL, NC, VA)

ESTIMATED POPULATION TREND 1966–2019: Not available

SIZE: Length 5 inches; wingspan 7 inches

The Saltmarsh Sparrow has a triangular orange patch
on the side of the face that contrasts with its white
throat and chest with chocolate-colored streaks.

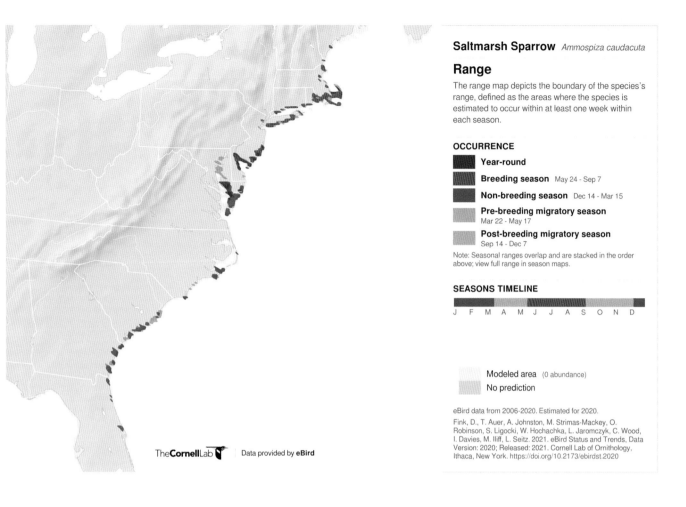

Saltmarsh Sparrow *Ammospiza caudacuta*

Range

The range map depicts the boundary of the species's range, defined as the areas where the species is estimated to occur within at least one week within each season.

OCCURRENCE

- Year-round
- Breeding season   May 24 - Sep 7
- Non-breeding season   Dec 14 - Mar 15
- Pre-breeding migratory season   Mar 22 - May 17
- Post-breeding migratory season   Sep 14 - Dec 7

Note: Seasonal ranges overlap and are stacked in the order above; view full range in season maps.

SEASONS TIMELINE

J F M A M J J A S O N D

Modeled area   (0 abundance)
No prediction

eBird data from 2006-2020. Estimated for 2020.

Fink, D., T. Auer, A. Johnston, M. Strimas-Mackey, O. Robinson, S. Ligocki, W. Hochachka, L. Jaromczyk, C. Wood, I. Davies, M. Iliff, L. Seitz. 2021. eBird Status and Trends, Data Version: 2020; Released: 2021. Cornell Lab of Ornithology, Ithaca, New York. https://doi.org/10.2173/ebirdst.2020

TheCornellLab | Data provided by **eBird**

Species Account   The Saltmarsh Sparrow inhabits coastal marshes in small numbers and has a patchy distribution along much of the Atlantic coast of the United States. In fact, it is the only passerine species in North America that is an obligate tidal marsh specialist (i.e., it lives only in coastal marshes). Like a golden-brown mouse, these sparrows run or walk on or near the ground, often moving under dense mats of salt meadow vegetation before popping up to scan their surroundings from a low perch in the grass. A heavy thatch layer appears to be an important habitat requirement, and some birds have been known to create or use tunnels through the thick vegetation, especially in the vicinity of the nest. Nests are often placed low to the ground, just above the normal high-tide line, making them vulnerable to flooding during storms.

Unlike many other birds, male Saltmarsh Sparrows typically do not sing to advertise their location or defend a specific territory in the marsh. Rather, males roam the habitat looking for receptive females. Once mating takes place, the female alone constructs the nest and cares for the young. Renesting occurs if a storm or predator destroys the first nest.

Until recently, the Saltmarsh Sparrow and the Nelson's Sparrow were considered the same species, and together they were known as the Sharp-tailed Sparrow, referring to the somewhat pointed tip of the notched, relatively short tail. In 1995 the Sharp-tailed Sparrow was split into two species: the Saltmarsh Sparrow, which nests along the coast, and the Nelson's Sparrow, which nests in the northern Great Plains, along the coast of Hudson Bay, and in portions of Maine and Nova Scotia. Genetic data, different vocalizations, and slight plumage variations led to the split. There is a small area of geographic overlap between the two species during the breeding season in southern coastal Maine, and some hybridization occurs there.

Although Breeding Bird Survey data are not available for the Saltmarsh Sparrow, range-wide surveys cited by the Cornell Lab of Ornithology indicate that the global population declined an average of 9 percent per year between 1998 and 2012, leading to a population decline of 75 percent in this short time frame. Certainly the loss of coastal marsh habitat has played an important role in the decline of this species. Other important factors may include invasive species such as phragmites, which degrade the remaining habitat, and rising sea levels tied to climate change, leading to increasingly higher high-tide levels. Pollutants may also play a role, as Saltmarsh Sparrows contain unusually high amounts of mercury compared to other sparrow species, which may influence reproductive success or survival rates.

It is estimated that only fifty thousand to sixty thousand Saltmarsh Sparrows remain today. If the current rate of population decline goes unchecked and sea levels continue to rise, inundating coastal habitats, multiple models predict that these birds could face extinction by the middle of the century. Additional study is required to better understand the habitat needs of these birds, and significant efforts should be made to stabilize and protect large coastal marshes along the Atlantic shoreline.

Identification The Saltmarsh Sparrow is heavily streaked with chocolate-colored bars across the breast and down the flanks. A bright orange triangle on the face contrasts with the much lighter orange or white of the throat and chest. The very similar Nelson's Sparrow also has an orange triangle on the face, but it blends seamlessly into the similar orange coloration of the throat and breast. Seaside Sparrows commonly occur in the same habitat as Saltmarsh Sparrows, but they are bigger and more powerfully built, are usually a darker slate color with yellow near the eyes, and lack the orange facial triangle.

Vocalizations Saltmarsh Sparrows are silent much of the time. Unlike most passerines, the males do not sing territorially, but they do give short bursts of "whisper song" on a single perch or sometimes longer song while in flight or on a series of perches. The whisper song consists of a succession of high-pitched, unmusical squeaks and chirps with a reedy tone. Females do not sing, but they do give a variety of short call notes.

Nesting The nest structure is built by the female and is placed near but not on the ground, just above the normal high-tide level. The cup-like structure is sometimes augmented with a partial canopy or roof. Three to five whitish eggs, usually heavily speckled with reddish-brown spots, are laid. The young hatch within twelve days and leave the nest about ten days after hatching.

The tail is deeply notched, giving rise to the Saltmarsh Sparrow's former name: Sharp-tailed Sparrow.

The Saltmarsh Sparrow runs through tunnels in the soft grasses of the salt marsh and suddenly pops out to survey its surroundings.

Henslow's Sparrow
(*Centronyx henslowii*)

CONSERVATION CONCERN SCORE: 14 (High)

OTHER DESIGNATIONS: American Bird Conservancy watch list (Red),
2021 USFWS Birds of Conservation Concern, 2016 Partners in Flight Species of
Continental Concern (Yellow), State Protected (AL), State Special Concern
(GA, MD, NC), State Threatened (VA), Species of Greatest Conservation Need
(FL, KY, MS, PA, SC, TN, WV)

ESTIMATED POPULATION TREND 1966–2019: −65%

SIZE: Length 5 inches; wingspan 7 inches

The Henslow's Sparrow has a subtle beauty, with hints of brown,
red, and green making it more colorful than many other sparrows.

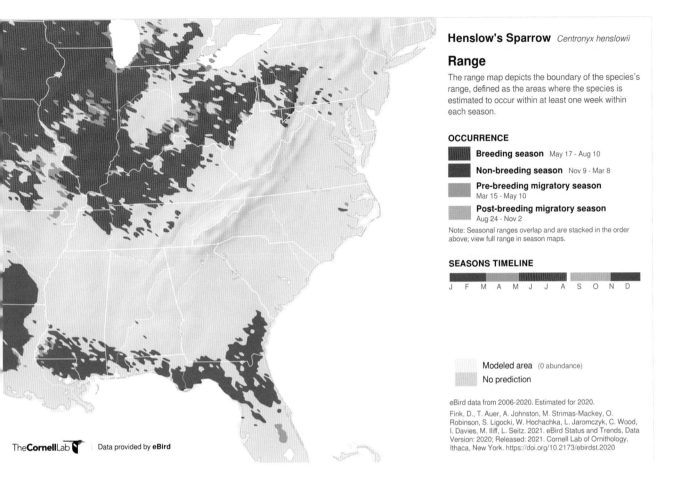

Henslow's Sparrow *Centronyx henslowii*

Range

The range map depicts the boundary of the species's range, defined as the areas where the species is estimated to occur within at least one week within each season.

OCCURRENCE

■ **Breeding season** May 17 - Aug 10

■ **Non-breeding season** Nov 9 - Mar 8

■ **Pre-breeding migratory season** Mar 15 - May 10

■ **Post-breeding migratory season** Aug 24 - Nov 2

Note: Seasonal ranges overlap and are stacked in the order above; view full range in season maps.

SEASONS TIMELINE

J F M A M J J A S O N D

Modeled area (0 abundance)

No prediction

eBird data from 2006-2020. Estimated for 2020.

Fink, D., T. Auer, A. Johnston, M. Strimas-Mackey, O. Robinson, S. Ligocki, W. Hochachka, L. Jaromczyk, C. Wood, I. Davies, M. Iliff, L. Seitz. 2021. eBird Status and Trends, Data Version: 2020; Released: 2021. Cornell Lab of Ornithology, Ithaca, New York. https://doi.org/10.2173/ebirdst.2020

TheCornellLab | Data provided by eBird

Species Account It is not easy to observe Henslow's Sparrows. The species has declined to the point where it is now largely limited to just a few prairies and wet meadows of sufficient size and quality to support a population. Making observation even more difficult, the birds are skulkers and prefer to stay close to the ground in thick vegetation. In fact, on the species' wintering grounds along the Gulf coast, observers attempting to survey the population spread out in a line and beat the grass with poles or sticks, hoping to catch a fleeting glimpse of a bird in flight. Undoubtedly the best time to see a Henslow's Sparrow is during the breeding season, when the males occasionally fly to the top of weed stalks to deliver their song, but even then, this bird can be hard to spot. The song is a short, weak *ss-lick* that sounds more like an insect than a bird. Its short duration, high pitch, and low volume all combine to make it difficult to judge the direction of and distance to the singer. However, when perseverance and luck pay off, the reward of seeing a Henslow's Sparrow is well worth the effort. Its muted shades of olive, buff, and rust have a subtle beauty when seen in good light, and the birds' declining numbers make such a sighting even more special.

Henslow's Sparrows once bred across the tallgrass prairies of the Midwest and in appropriate habitat along the Atlantic seaboard. Although the southeastern states provide critical winter habitat for Henslow's Sparrows, in the region covered by this book, they breed only from Pennsylvania down through Maryland, West Virginia, Kentucky, and Tennessee, in addition to a small, isolated breeding population in North Carolina as of this writing. The bird was first described by John James Audubon from a specimen collected in Kentucky in 1820. Audubon named the bird in honor of John Stevens Henslow, a friend and professor of botany at Cambridge University who had helped Audubon sell subscriptions to his *Birds of North America* during an earlier visit to England.

Appropriate management of the Henslow's Sparrow's remaining breeding habitat is the subject of debate. Prescribed fire as a management tool for the birds' preferred wet grassland habitat is often disparaged because it removes the layer of dead grasses from previous growing seasons. After a fire, Henslow's Sparrows are often absent for several years before the habitat becomes sufficiently lush again. However, in the absence of fire, grasslands quickly become too brushy and are abandoned altogether. In the past, when the birds had access to swaths of prairie covering hundreds or even thousands of square miles, a mosaic of suitable microhabitats was common, including various stands of grasses that had burned at different intervals. In that scenario, some habitat was always available to meet the Henslow's Sparrow's needs. Today, with prairie habitat present only in small, isolated chunks, it is much more difficult to replicate that mosaic.

Identification The bird appears large billed and large headed for its body size. The face has an olive cast that extends to the nape of the neck, a white eye ring, and a brown streak behind the eye. Fine, dark streaks are confined to a band across the breast and extending down the flanks. The back appears scaly because of the white edges of some feathers, and the color is a blend of rufous, black, and tan. Males and females are identical. In the Southeast, the Henslow's Sparrow is most likely to be confused with the Grasshopper Sparrow, which lacks the black streaks on the breast and the olive cast of the face.

Vocalizations The song is a weak, insect-like, two-part *ss-lick* that has been described as a feeble hiccup. Calls consist of a high, sharp *tsik* note.

Nesting The nest is a deep cup of grasses and leaves that is sometimes lined with hair. It is built within twenty inches of the ground, often at the base of a tuft of grass. The female builds the nest alone, which takes four to six days. Clutches typically include three to five glossy white eggs heavily marked with reddish brown. Several pairs may nest in close proximity in a loose colony.

Although the song of the Henslow's Sparrow is very weak, it is the best way to locate these birds, which love to skulk in the grasses.

Golden-winged Warbler
(*Vermivora chrysoptera*)

CONSERVATION CONCERN SCORE: 16 (High)

OTHER DESIGNATIONS: American Bird Conservancy watch list (Red),
2021 USFWS Birds of Conservation Concern, 2016 Partners in Flight Species of
Continental Concern (Red), 2021 Audubon Priority Birds list, State Special Concern
(GA, MD, NC), Species of Greatest Conservation Need (FL, KY, PA, SC, TN, VA, WV)

ESTIMATED POPULATION TREND 1966–2019: −63%

SIZE: Length 5 inches; wingspan 8 inches

Golden-winged Warblers prefer shrubby wetlands—a habitat
that is disappearing across the American landscape.

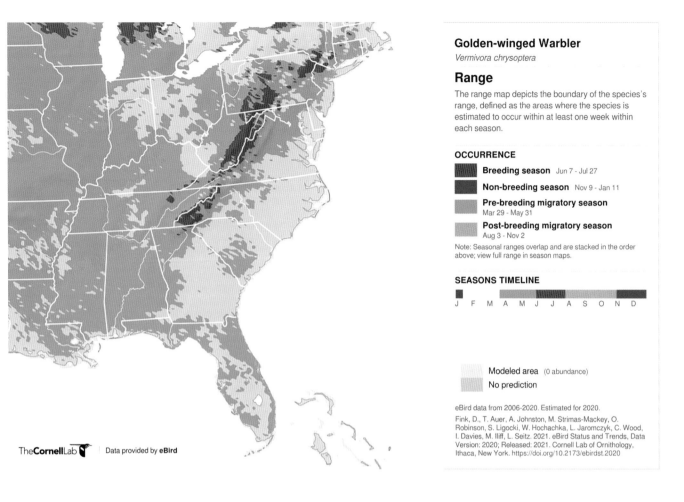

Golden-winged Warbler

Vermivora chrysoptera

Range

The range map depicts the boundary of the species's range, defined as the areas where the species is estimated to occur within at least one week within each season.

OCCURRENCE

■	**Breeding season** Jun 7 - Jul 27
■	**Non-breeding season** Nov 9 - Jan 11
■	**Pre-breeding migratory season** Mar 29 - May 31
■	**Post-breeding migratory season** Aug 3 - Nov 2

Note: Seasonal ranges overlap and are stacked in the order above; view full range in season maps.

SEASONS TIMELINE

J F M A M J J A S O N D

Modeled area (0 abundance)

No prediction

eBird data from 2006-2020. Estimated for 2020.

Fink, D., T. Auer, A. Johnston, M. Strimas-Mackey, O. Robinson, S. Ligocki, W. Hochachka, L. Jaromczyk, C. Wood, I. Davies, M. Iliff, L. Seitz. 2021. eBird Status and Trends, Data Version: 2020; Released: 2021. Cornell Lab of Ornithology, Ithaca, New York. https://doi.org/10.2173/ebirdst.2020

TheCornellLab | Data provided by **eBird**

Species Account The Golden-winged Warbler is a sought-after species, with throngs of binocular-toting birders jamming the boardwalks and paths at popular migrant traps whenever one is seen during spring migration. In addition to its rarity, its beauty is sure to attract a crowd of enthusiastic observers. The Golden-winged Warbler has subtle, charcoal-gray plumage highlighted by brilliant golden yellow on the crown and wings. Its black face and throat patch, along with its habit of hanging upside down and probing rolled leaves for insects, might bring to mind a chickadee, but that thought quickly vanishes with a flash of gold from the crown or wing. The Golden-winged Warbler has been in steep decline for decades, but despite its small population, it has not been federally listed.

The reasons for the decline of the Golden-winged Warbler are common to many species described in this book. Loss of habitat is surely one of the most important. Like many birds with rapidly declining populations, Golden-winged Warblers need early successional habitat. In the past, this patchy, shrubby habitat intermixed with more mature deciduous woodlands was associated with recent wildfires, beaver-flooded areas, abandoned farmlands, and shrubby wetlands. As these habitats have grown into mature forests or been drained (in the case of wetlands), less appropriate habitat is left to support Golden-winged Warblers. In some cases, remaining wetlands have been taken over by nonnative invasive plant species, altering the habitat and making it unsuitable for nesting. Brown-headed Cowbirds have also played a role in reducing the Golden-winged Warbler population, with one study finding a 17 percent reduction in warbler productivity because of nest parasitism by cowbirds (Confer and Larkin 2003).

However, another reason for the decline of the Golden-winged Warbler is not as common or as obvious. Golden-winged Warblers are apparently being outcompeted by the closely related Blue-winged Warbler and absorbed through interbreeding with that species. Although they are also in decline, Blue-winged Warblers have shifted their range north and east, and Golden-winged Warblers appear to be losing ground to them. Studies have shown that once blue-wings invade an area, golden-wings disappear in as little as fifty years, partly because they are competing for similar resources. When the two species interbreed, they produce viable hybrid offspring known as Brewster's or Lawrence's Warblers, depending on their physical characteristics. As these hybrid offspring breed with members of the pure parent species, the two species' genetic makeups become increasingly blurred. Overall, however, more blue-winged genes are ending up in the golden-winged population. Today, only a small portion of the Golden-winged Warbler population (primarily in Manitoba, Canada) is genetically pure, and even birds that appear to be pure Golden-winged Warblers often have traces of Blue-winged Warbler genes. It is unclear why the Blue-winged Warbler's genetic makeup is more likely to remain intact while the Golden-winged Warbler's genetic purity is being lost. However, some studies have shown that Golden-winged Warblers are much more likely to breed with hybrid offspring, which could play a role in the overall gene flow between the species.

Recent surveys have shown a 98 percent decline in Golden-winged Warbler populations in the Adirondack Mountains of New York. In other areas, the Golden-winged Warbler has been completely lost as a breeding bird. Fortunately, a number of institutions and conservation organizations have come together to create the Golden-winged Warbler Working Group, which has devised a conservation plan that details what actions need to be taken to increase the species' chances of survival.

Identification The male Golden-winged Warbler has a jet-black cheek patch and bib, with contrasting white surrounding the black cheek. The breast and underparts are whitish gray, the back is steel gray, and there is a gold patch on the wings and crown. The female is similar but has dark charcoal (rather than black) patches on the throat and cheek and less intense gold on the crown. The Chestnut-sided Warbler, which breeds in similar habitat, also has a golden crown but lacks the bold black bib and golden wing patch of the Golden-winged Warbler. The hybrid Brewster's Warbler has the overall coloration of the Golden-winged Warbler and the facial pattern of the Blue-winged Warbler, while the hybrid Lawrence's Warbler has the overall coloration of the Blue-winged Warbler and the facial pattern of the Golden-winged Warbler. Some variation is possible.

Vocalizations The male gives a slow, buzzy *bee bzzzz bzzzz*. The first note is higher than subsequent notes.

Nesting Golden-winged Warblers nest on the ground and lay three to six pale eggs that are lightly marked with fine splotches or streaks toward the larger end. The nest is often situated at the base of a plant stem. Both parents feed the young until fledging takes place at approximately ten days of age.

The male Golden-winged Warbler has distinctive black facial markings and brilliant yellow on the crown and wings, making him very dashing indeed.

Blue-winged Warbler
(*Vermivora cyanoptera*)

CONSERVATION CONCERN SCORE: 13 (Moderate)

OTHER DESIGNATIONS: American Bird Conservancy watch list (Yellow),
  Species of Greatest Conservation Need (KY, MD, PA, SC, TN, WV)

ESTIMATED POPULATION TREND 1966–2019: −32%

SIZE: Length 5 inches; wingspan 8 inches

The male Blue-winged Warbler is golden overall and generally similar
to the Prothonotary Warbler, but he has two bright white wing bars
and a dark line through the eye that almost looks like a mask.

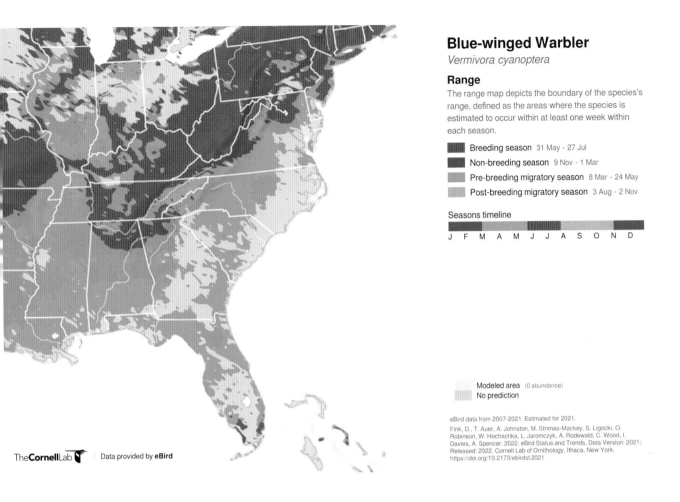

Blue-winged Warbler
Vermivora cyanoptera

Range

The range map depicts the boundary of the species's range, defined as the areas where the species is estimated to occur within at least one week within each season.

■ Breeding season 31 May - 27 Jul
■ Non-breeding season 9 Nov - 1 Mar
■ Pre-breeding migratory season 8 Mar - 24 May
■ Post-breeding migratory season 3 Aug - 2 Nov

Seasons timeline

J F M A M J J A S O N D

Modeled area (0 abundance)
No prediction

eBird data from 2007-2021. Estimated for 2021.
Fink, D., T. Auer, A. Johnston, M. Strimas-Mackey, S. Ligocki, O. Robinson, W. Hochachka, L. Jaromczyk, A. Rodewald, C. Wood, I. Davies, A. Spencer. 2022. eBird Status and Trends, Data Version: 2021; Released: 2022. Cornell Lab of Ornithology, Ithaca, New York. https://doi.org/10.2173/ebirdst.2021

TheCornellLab | Data provided by eBird

Species Account The Blue-winged Warbler inhabits scrubby thickets, forest openings, shrubby swamps, old fields, recovering clear-cuts, and other areas with young forest vegetation. Prior to the early 1800s, this species was likely confined to the Ozarks and the savannas of Kentucky and Tennessee. Once settlers began the extensive cutting of mature deciduous forests, the Blue-winged Warbler enjoyed a significant range expansion to much of the Midwest and the Northeast. This species reached Ohio and southern Michigan by the early 1900s and central New York by the 1940s. As it advanced north and increased in portions of the Appalachians where it had previously been absent, the Blue-winged Warbler began to compete with and displace the Golden-winged Warbler, which prefers similar shrubland habitat. With the invasion of the Blue-winged Warbler, the Golden-winged Warbler retreated farther north, and its population declined precipitously. This displacement and decline are evident in the fall migration records from Cape May, New Jersey: In the 1920s, Blue-winged Warblers outnumbered Golden-winged Warblers at Cape May by two to one. Today, Blue-winged Warblers outnumber Golden-winged Warblers by thirty to one. For

The Blue-winged Warbler is an active forager, often probing clusters of dead leaves or hanging upside down to search for caterpillars or other food items hidden in the vegetation.

additional details about the relationship between these species, see the chapter on the Golden-winged Warbler.

The Blue-winged Warbler can be found in shrubs, in saplings, or at the forest edges, working the outer branches of trees and sometimes hanging upside down or hovering as it attempts to locate prey. This species often probes clumps of dead leaves, buds, or flowers looking for caterpillars, spiders, ants, and beetles. The birds may also forage closer to the ground, where they take grasshoppers and crickets.

Unfortunately, the Brown-headed Cowbird often lays its eggs in a Blue-winged Warbler's nest. The warbler parents preferentially feed the larger cowbird chicks over their own young, resulting in reduced growth rates for young warblers, increased mortality for warbler chicks, and less overall reproductive success for Blue-winged Warblers. This is especially true in fragmented habitats, where cowbird nest parasitism may be as high as 67 percent. Conversely, in a West Virginia study of large areas of intact habitat, cowbird parasitism was recorded in only one of 212 Blue-winged Warbler nests (Canterbury, Kotesovec, and Catuzza 1995).

Although cowbirds definitely impact Blue-winged Warblers, the primary cause of the species' recent decline is loss of habitat. Over time, the pace of timber clearing in the East has slowed, and over the last few decades, the average age of forests has increased. The resulting loss of young forests and shrublands has adversely impacted many early successional birds, including the Blue-winged Warbler. Despite a massive population increase during the 1800s and early 1900s, Breeding Bird Survey data indicate a continental decline of more than 30 percent since the late 1960s. However, this decline is not nearly as severe as that experienced by the Golden-winged Warbler.

Identification The adult male Blue-winged Warbler is bright lemon yellow across the head, throat, and belly. The back is green, and the wings are slaty blue with bold white wing bars. A small black line extends from the base of the beak past the eye. Female birds may be less boldly marked, although some are indistinguishable from males.

Vocalizations Males call in the spring, often from a conspicuous perch. The song is a two-syllable, reedy *bee buzzzz,* with the second note slightly longer and at a lower pitch than the first.

Nesting Most nests are located in areas with dense saplings near forest edges but at least ninety feet from large blocks of mature woods. A West Virginia study found an average of 112 small saplings within forty feet of nests (Will 1986). Females choose the nest location, which is usually on or near the ground. Nests are made of grasses and often include dead leaves and bark from grapevines. A typical clutch consists of four or five white eggs with light-brown speckles toward the larger end.

Prothonotary Warbler
(*Protonotaria citrea*)

CONSERVATION CONCERN SCORE: 14 (High)

OTHER DESIGNATIONS: American Bird Conservancy watch list (Yellow),
2021 USFWS Birds of Conservation Concern, 2021 Audubon Priority Birds list,
Species of Greatest Conservation Need (FL, GA, KY, MD, MS, NC, PA, SC, TN, WV)

ESTIMATED POPULATION TREND 1966–2019: −32%

SIZE: Length 6 inches; wingspan 9 inches

The sight of a male Prothonotary Warbler's glowing yellow
plumage in a shady forested wetland makes a lasting impression.

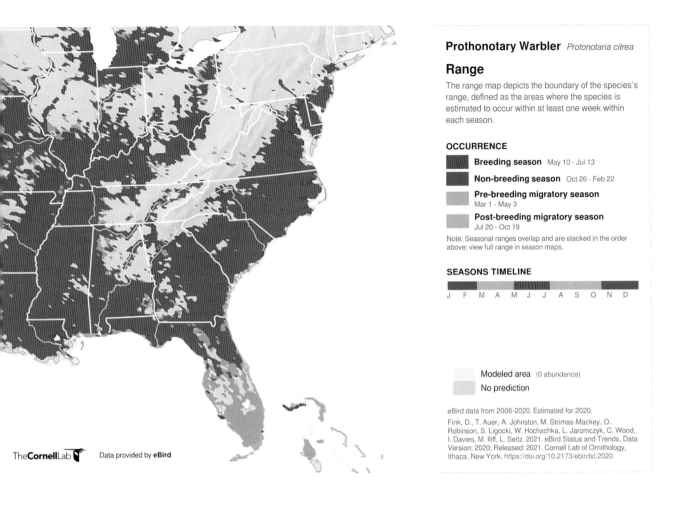

Prothonotary Warbler *Protonotaria citrea*

Range

The range map depicts the boundary of the species's range, defined as the areas where the species is estimated to occur within at least one week within each season.

OCCURRENCE

- **Breeding season** May 10 - Jul 13
- **Non-breeding season** Oct 26 - Feb 22
- **Pre-breeding migratory season** Mar 1 - May 3
- **Post-breeding migratory season** Jul 20 - Oct 19

Note: Seasonal ranges overlap and are stacked in the order above; view full range in season maps.

SEASONS TIMELINE

J F M A M J J A S O N D

- Modeled area (0 abundance)
- No prediction

eBird data from 2006-2020. Estimated for 2020.

Fink, D., T. Auer, A. Johnston, M. Strimas-Mackey, O. Robinson, S. Ligocki, W. Hochachka, L. Jaromczyk, C. Wood, I. Davies, M. Iliff, L. Seitz. 2021. eBird Status and Trends, Data Version: 2020; Released: 2021. Cornell Lab of Ornithology, Ithaca, New York. https://doi.org/10.2173/ebirdst.2020

TheCornellLab | Data provided by eBird

Species Account Prothonotary Warblers reside in the swampy forests and riparian woodlands of the Southeast. A brilliant yellow male suddenly illuminated by a shaft of sunlight in its dark, shaded habitat is a memorable sight. The name *prothonotary* refers to scribes of the Roman Catholic Church who wore yellow hoods. Golden Swamp Warbler has been suggested as another common name for this species and fits the bird equally well. The Prothonotary Warbler may be most numerous in the wooded floodplains associated with the Mississippi River Valley, as well as in the swampy woods of coastal Virginia and the Carolinas.

Males are the first to return from their wintering grounds in the lowland forests and mangroves of coastal Mexico and Central and South America, often arriving along the Gulf coast during the last week of March. However, the majority of birds return to their nesting grounds in the Appalachians and southeastern United States in April. The males quickly start to search for suitable nesting cavities—the Prothonotary is the only cavity-nesting warbler species east of the Mississippi. Although they primarily use the old nest sites of other species, especially the Downy Woodpecker, they occasionally excavate their own nest cavities using their relatively hefty beaks, which are unlike those of other warblers. Still, this is possible only if they can find suitable locations in soft, decayed wood. The male warbler starts carrying nesting material to at least one cavity, which he shows off to prospective females. Once a mate is found, the female finishes constructing the nest.

Nesting cavities are typically located over water and five to ten feet high in a dead tree or dead branch. However, they can be as low as two to three feet above the waterline, which makes them susceptible to flooding. Parasitism by Brown-headed Cowbirds has been reported, and some studies indicate a surprising number of eggs destroyed by House Wrens, which may compete for similar nest sites. However, the major cause of population decline for the Prothonotary Warbler is loss of habitat due to the draining, filling, and clearing of its preferred habitat.

Prothonotary Warblers are most often encountered by canoers and kayakers, or they may be seen along a boardwalk that traverses flooded forest. They are frequently located by their ringing *sweet sweet sweet sweet sweet* song or their metallic-sounding alarm calls. The birds are active foragers and can be seen landing on half-submerged mossy logs, peeking into crevices, or perching on the sides of tree trunks. They seem rather tame and sometimes land quite near those who sit quietly and wait.

Identification Prothonotary Warblers are brilliant golden yellow on the head and chest, with unmarked bluish-gray wings, a greenish back, and white underparts. They have a relatively heavy black beak and black feet. Females are similar to males but less brilliantly marked. White spots on the tail may be visible when it is spread during flight. The Blue-winged Warbler is similar but has a small black mask between the beak and eyes and two white wing bars.

Vocalizations The song is a ringing, metallic *sweet* note repeated seven to ten times in rapid succession. The call note is a metallic-sounding *chip*.

Nesting Prothonotary Warblers often nest in snags or dead tree limbs along rivers, lakes, sloughs, or other bodies of water. Some birds reportedly use nest boxes and other human-made structures for nesting, including posts, buildings, empty tin cans, mailboxes, and even an old pail on the porch of a cabin near a stream. (Plans for making a Prothonotary Warbler nest box can be found at https://www.tn.gov/twra /wildlife/woodworking-for-wildlife/prothonotary-warbler-nest-box.html.) The nest cup is about two inches wide and is made of roots, bark, moss, lichen, and other plant material. A clutch consists of three to seven white eggs with brownish spots, but productivity is thought to be low. A recent study found that in only 28 percent of the 178 nests examined did at least one chick survive to fledging.

The female Prothonotary Warbler is similar to the male
but a shade or two less intense in coloration.

Prothonotary Warblers are rather tame, and it is not unusual for one to approach at eye level if you sit quietly.

Swainson's Warbler
(*Lymnothlypis swainsonii*)

CONSERVATION CONCERN SCORE: 13 (Moderate)

OTHER DESIGNATIONS: American Bird Conservancy watch list (Yellow),
2021 Audubon Priority Birds list, In Need of Management (TN), State Special
Concern (GA), State Endangered (MD), Species of Greatest Conservation Need
(KY, MS, NC, SC, VA, WV)

ESTIMATED POPULATION TREND 1966–2019: +68%

SIZE: Length 6 inches; wingspan 9 inches

Swainson's Warblers are typically found in riparian forests
of the South or rhododendron thickets in the Appalachians.
They usually remain on or near the ground, where their muted
plumage helps hide them. (Photograph by Caben Williams)

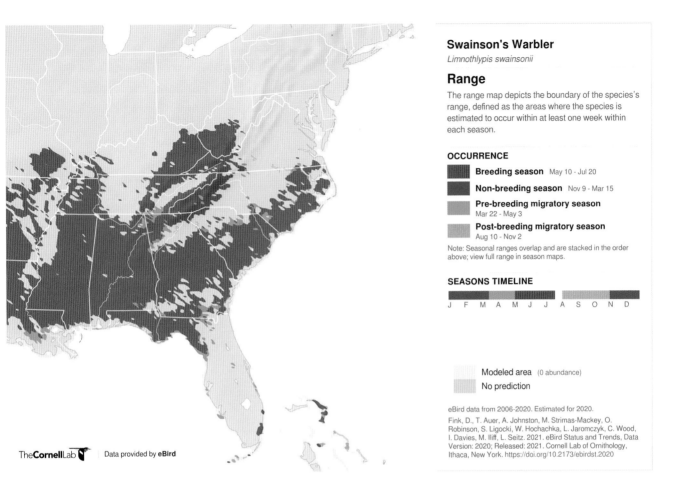

Swainson's Warbler
Limnothlypis swainsonii

Range

The range map depicts the boundary of the species's range, defined as the areas where the species is estimated to occur within at least one week within each season.

OCCURRENCE

- **Breeding season** May 10 - Jul 20
- **Non-breeding season** Nov 9 - Mar 15
- **Pre-breeding migratory season** Mar 22 - May 3
- **Post-breeding migratory season** Aug 10 - Nov 2

Note: Seasonal ranges overlap and are stacked in the order above; view full range in season maps.

SEASONS TIMELINE

J F M A M J J A S O N D

Modeled area (0 abundance)

No prediction

eBird data from 2006-2020. Estimated for 2020.

Fink, D., T. Auer, A. Johnston, M. Strimas-Mackey, O. Robinson, S. Ligocki, W. Hochachka, L. Jaromczyk, C. Wood, I. Davies, M. Iliff, L. Seitz. 2021. eBird Status and Trends, Data Version: 2020; Released: 2021. Cornell Lab of Ornithology, Ithaca, New York. https://doi.org/10.2173/ebirdst.2020

TheCornellLab | Data provided by eBird

Species Account Fairly secretive, plainly colored, and preferring a habitat that includes dense undergrowth, the Swainson's Warbler is a difficult bird to get to know and is likely heard much more often than it is seen. With a patchy distribution throughout the southern Appalachians and the Southeast, the Swainson's Warbler can be found in drier bottomland hardwood forests, in rhododendron thickets in the Appalachians, and in young pine plantations and dense stands of live oak with greenbrier thickets along the coastal plain. Historically, the species was common in the bamboo-like stands of native giant cane along southern rivers, but much of this habitat has been completely destroyed. Overall, the species prefers a large forest block with moist soils, a full canopy cover with some gaps that admit enough light to support a dense undergrowth of shrubs or vines, and abundant leaf litter.

John Abbot, an artist from Georgia, created drawings and watercolors of this species in 1801—decades before it was officially described by John James Audubon based on a specimen he received from a friend in South Carolina in 1833. The first nest of this species was not found until 1885. The scientific name for this species, *Limnothlypis,*

The song of
the Swainson's
Warbler may be
mistaken for that
of the Louisiana
Waterthrush, which
is sometimes found
in the same habitat.

means "marsh finch." This name was chosen because Swainson's Warblers are often found near swamps or in bottomlands along rivers. However, they are not tolerant of long-term flooding and tend to avoid areas that are inundated for much of the year.

Swainson's Warblers spend much of their time on or near the ground. They frequently forage in the leaf litter by turning over leaves and searching for prey. The birds may also insert their bills into rolled leaves, looking for insects. Common food items include spiders, crickets, beetles, centipedes, and various larvae. A study from South Carolina indicates that larval and adult spiders, larval butterflies and moths, and larval bees and wasps are preferred (Savage et al. 2010). On its wintering grounds in the Caribbean, the species also consumes small lizards and geckos, in addition to the more common beetles, spiders, and ants.

Although Breeding Bird Survey data indicate a rapidly increasing population since the 1970s, caution is advised in interpreting these data. Relatively few BBS routes overlap with good Swainson's Warbler habitat, so the trends are based on a fairly small number of transects. Certainly, the overall population of this species has declined since the early 1800s, as 98 percent of the native giant cane thickets have been lost and large tracts of bottomland forests have been fragmented. Overall, 55 to 80 percent of the original bottomland hardwood forests of the southeastern United States are gone, representing a significant loss of this species' preferred habitat. In addition, Brown-headed Cowbirds parasitize this species, negatively affecting reproduction. Recent estimates indicate that there are 140,000 Swainson's Warblers left, which is actually a small population compared with that of most other eastern warblers, justifying its inclusion in this book.

Identification The Swainson's Warbler has a plain brown back with a slightly more reddish-brown cap. The belly is white and unstreaked, sometimes with a yellowish tinge. There is a strong white supercilium and a line through the eye. The legs are pink.

Vocalizations The song is reminiscent of that of a Louisiana Waterthrush and has been described as *whee-whee-whee-whip-poor-will.*

Nesting Males typically hold large territories. The nest is often placed in a thick growth of vines or cane within ten feet of the ground. The average nest height in Alabama was about five and a half feet (Summerour 2008). The nests of Swainson's Warblers are the largest of all the North American warblers—about the size of a Northern Cardinal nest—and are made of leaves, small roots, Spanish moss, and other materials in a loose cup. The female constructs the nest over the course of a few days. Three unmarked white eggs are typical, although some eggs have reddish-brown speckling around the large end.

Kentucky Warbler
(*Geothlypis formosa*)

CONSERVATION CONCERN SCORE: 14 (High)

OTHER DESIGNATIONS: American Bird Conservancy watch list (Yellow),
2021 USFWS Birds of Conservation Concern, 2016 Partners in Flight Species of
Continental Concern (Yellow), Species of Greatest Conservation Need (FL, KY,
MD, MS, NC, PA, SC, TN, VA, WV)

ESTIMATED POPULATION TREND 1966–2019: −32%

SIZE: Length 5 inches; wingspan 9 inches

The adult male Kentucky Warbler has extensive
black on the face and cap, a lemon-yellow body,
and plain brown on the back and upper tail.

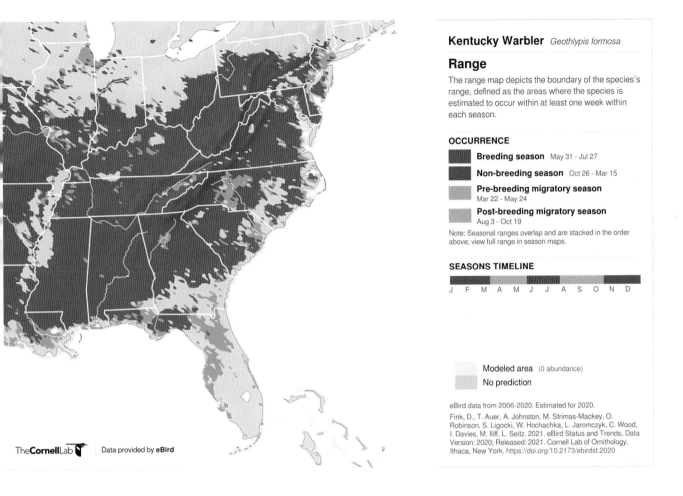

Kentucky Warbler *Geothlypis formosa*

Range

The range map depicts the boundary of the species's range, defined as the areas where the species is estimated to occur within at least one week within each season.

OCCURRENCE

	Breeding season May 31 - Jul 27
	Non-breeding season Oct 26 - Mar 15
	Pre-breeding migratory season Mar 22 - May 24
	Post-breeding migratory season Aug 3 - Oct 19

Note: Seasonal ranges overlap and are stacked in the order above; view full range in season maps.

SEASONS TIMELINE

J F M A M J J A S O N D

	Modeled area (0 abundance)
	No prediction

eBird data from 2006-2020. Estimated for 2020.

Fink, D., T. Auer, A. Johnston, M. Strimas-Mackey, O. Robinson, S. Ligocki, W. Hochachka, L. Jaromczyk, C. Wood, I. Davies, M. Iliff, L. Seitz. 2021. eBird Status and Trends, Data Version: 2020; Released: 2021. Cornell Lab of Ornithology, Ithaca, New York. https://doi.org/10.2173/ebirdst.2020

TheCornellLab | Data provided by **eBird**

Species Account Like the Mourning Warbler and Connecticut Warbler—which, until recently, were in the same genus—the Kentucky Warbler is at home on or near the ground. Kentucky Warblers hop on the ground and turn over leaf litter or inspect overhanging leaves to find their prey, which includes spiders, caterpillars, beetles, and ants. During the breeding season, Kentucky Warblers seek deep woods with a shaded, well-developed, brushy understory; they seem to prefer thickly vegetated ravines and well-shaded slopes. In winter the birds can be found in humid, forested lowlands in Mexico, Central America, and portions of the Caribbean, south to Colombia and Venezuela. Even in winter, a dense shrub layer in shaded forest habitat appears to be key, and this species is sensitive to forest fragmentation.

Across the Southeast, the Kentucky Warbler is still a relatively common bird in appropriate habitat, despite recent declines. However, the birds are easier to hear than to see because of their preference for rugged terrain and a thick understory. Even singing males can be hard to locate because they sometimes sit very still for minutes at a time when vocalizing, and it can be difficult to judge distance and direction. In addition,

This Kentucky Warbler, which has some black on the face and very
little black on the cap, is likely an adult female or a young male.

many Kentucky Warblers are misidentified because their loud, rolling *tur-dle tur-dle* is
mistaken for the similar song of the Carolina Wren.

Among warblers, the Kentucky is a relatively early spring migrant. The first birds
cross the Gulf of Mexico and arrive along the US Gulf coast as early as mid to late
March. Farther north in the Appalachians, the first spring arrivals appear in mid to
late April, with some birds not arriving until early May. Occasionally, birds overshoot
their normal breeding range and are recorded farther north in the spring. Singing
males may be present at these northern locations for a short time, but it is doubtful
that breeding occurs. Fall migration can begin as early as late July, although the main
migration takes place in September. By early October, most of the birds have left West
Virginia, Kentucky, and Tennessee, and by the end of the month, even those lingering
along the Gulf coast have departed for points farther south.

The Kentucky Warbler is a fairly frequent victim of nest parasitism by the Brown-
headed Cowbird. Lawns and other open spaces that fragment forests may provide
opportunities for the cowbirds to locate and lay their eggs in the nests of Kentucky
Warblers. A study of 250 Kentucky Warbler nests found that the rate of nest parasitism by cowbirds varied from 60 percent near a cowbird feeding area to only 3 percent
a mile or more away from a cowbird feeding site. The same study also found lower
parasitism rates in older forest stands than in younger forests. Deforestation of the

This juvenile Kentucky Warbler, most likely a female, has very little black anywhere on the head.

species' wintering grounds and deer overpopulation resulting in excessive browsing and thinner forest understories have been suggested as other reasons for the birds' recent decline.

Identification The Kentucky Warbler is a stocky, short-tailed bird with a plain, unmarked olive back and wings and lemon-yellow underparts. The long legs are bright pink. The male has a black cheek patch that extends down the side of the throat, as well as black on the crown that fades to gray toward the nape. A bright yellow line extends from the bill and wraps around the back of the eye. The female is similar, but with less black on the face and head. Similar species include the Common Yellowthroat, which is smaller, has a complete black mask (males), and is usually found in wetland habitats.

Vocalizations The male sings a ringing, rich *tur-dle tur-dle tur-dle* that is quite similar to the Carolina Wren's three-syllable *tea kettle tea kettle tea kettle*.

Nesting Nests are large, loose cups placed on the ground or just above it. The nest is constructed by both sexes and is made up of grass, bark, roots, and leaves. Four or five whitish eggs with reddish-brown speckles are typically laid and are incubated for nearly two weeks before hatching.

Cerulean Warbler
(*Setophaga cerulea*)

CONSERVATION CONCERN SCORE: 15 (High)

OTHER DESIGNATIONS: American Bird Conservancy watch list (Yellow),
    2021 USFWS Birds of Conservation Concern, 2016 Partners in Flight Species of
    Continental Concern (Yellow), 2021 Audubon Priority Birds list, State Protected (AL),
    In Need of Management (TN), State Special Concern (GA, NC),
    Species of Greatest Conservation Need (FL, GA, KY, MD, MS, PA, SC, VA, WV)

ESTIMATED POPULATION TREND 1966–2019: –65%

SIZE: Length 5 inches; wingspan 8 inches

The Cerulean Warbler, which inhabits large tracts of mature
forest, has lost significant amounts of habitat both in its
breeding grounds in the eastern United States and in South
America, where it spends the nonbreeding season.

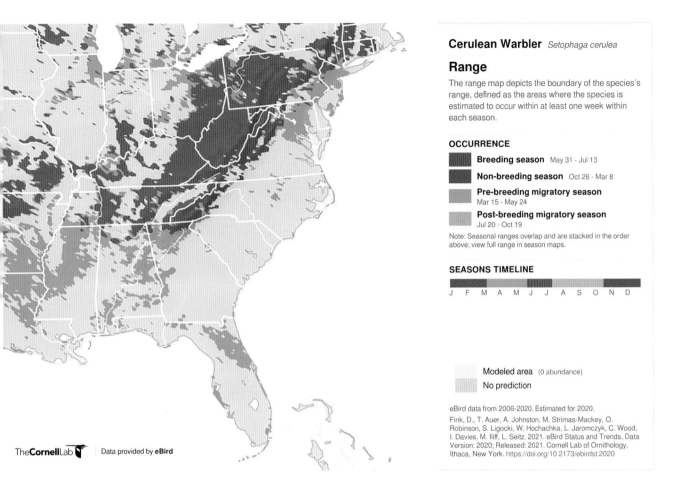

Cerulean Warbler *Setophaga cerulea*

Range

The range map depicts the boundary of the species's range, defined as the areas where the species is estimated to occur within at least one week within each season.

OCCURRENCE

- **Breeding season** May 31 - Jul 13
- **Non-breeding season** Oct 26 - Mar 8
- **Pre-breeding migratory season** Mar 15 - May 24
- **Post-breeding migratory season** Jul 20 - Oct 19

Note: Seasonal ranges overlap and are stacked in the order above; view full range in season maps.

SEASONS TIMELINE

J F M A M J J A S O N D

Modeled area (0 abundance)

No prediction

eBird data from 2006-2020. Estimated for 2020.

Fink, D., T. Auer, A. Johnston, M. Strimas-Mackey, O. Robinson, S. Ligocki, W. Hochachka, L. Jaromczyk, C. Wood, I. Davies, M. Iliff, L. Seitz. 2021. eBird Status and Trends, Data Version: 2020; Released: 2021. Cornell Lab of Ornithology, Ithaca, New York. https://doi.org/10.2173/ebirdst.2020

TheCornellLab | Data provided by **eBird**

Species Account Unlike many warblers of the eastern United States, the male Cerulean Warbler has no yellow markings adorning his plumage. Instead, the bird lives up to its name and sports a brilliant shade of cerulean blue on its head and back. This sky-blue coloration seems appropriate for a bird that spends so much time at the tops of large, mature trees, forcing observers to stare up into the canopy as they try to catch a glimpse of this tiny warbler. Although lacking the brilliant blue coloration of the males, females have uniquely beautiful plumage, with aqua or sea-green heads and upper backs.

Despite being studied fairly extensively, this species' exact habitat requirements remain debatable, although the consensus is that Cerulean Warblers require large, intact blocks of mature deciduous forest. Many sources agree on a minimum size of fifty to seventy-five acres of mature forest, with blocks of at least six hundred contiguous acres being preferred. Prime habitat for these birds often includes steep topography, and sites with streams seem to be favored. However, there is some evidence that relatively small breaks in the forest canopy might be important. These openings may be created by storms or by the existence of shrubby wetlands within the forest matrix, or they

The female Cerulean Warbler's soft aqua tones are unlike those of any other species in the United States.

may be deliberately created through different types of timber harvesting practices. Whatever their source, it seems that the structure provided by these small pockets of younger forest within a large matrix of mature deciduous forest, with a relatively open understory, may be even more critical than tree species or soil type. In fact, oak-hickory– and maple-dominated forests have both reportedly been used for nesting, as well as wet riparian forests and drier upland sites.

The Cerulean Warbler is one of the most imperiled migrant songbirds in the country and has experienced a steep population decline over the last fifty years. Differing slightly from Breeding Bird Survey data, Partners in Flight estimates that the population has dropped more than 70 percent since 1970 and predicts an additional 50 percent decline in the next few decades. Habitat loss is likely the biggest cause, resulting in fewer large forest blocks available for nesting. In the United States, Appalachia is incredibly important for Cerulean Warblers, with an estimated 80 percent of nesting taking place in the region. Loss of forested habitat in the Appalachians due to mineral extraction, agriculture, and urbanization has certainly taken a toll, but loss of forest habitat in the Andes Mountains of northern South America may be the biggest driver of population decline going forward, as more of the birds' wintering habitat is converted to pasture for livestock and the planting of crops such as rice, cocoa, and coffee. To date, more than 60 percent of the Cerulean Warbler's wintering habitat in South America may have been destroyed. However, the American Bird Conservancy and other groups have contributed to the establishment of forest reserves in the area, as well as the creation of conservation easements to protect the remaining forests. Reforestation projects have also created corridors of habitat between these reserves. More than two

Often found in the treetops, the male Cerulean Warbler's sky-blue coloration fits his preference for being high in the canopy.

hundred private landowners participated in a recent effort to plant shade trees on more than three thousand acres of coffee and cocoa farms to provide better habitat corridors between reserves in Colombia. Efforts like these may help turn the tide for this beautiful warbler of the treetops.

Identification The male Cerulean Warbler is nearly unmistakable, with a sky-blue head and back, a white belly, a blackish necklace across the chest, and dark streaking down the flanks and on the back. The female is aqua to green on the crown and upper back, with whitish or yellowish underparts; there is faint, dark streaking along the sides of the breast and flanks and a white supercilium streak. Both sexes have two strong white wing bars.

Vocalizations The song is a rapid series of short buzzy notes, followed by a longer, higher-pitched trill.

Nesting The nest is constructed in the mid to upper canopy of a deciduous tree—from as low as fifteen feet to as high as ninety feet. Oak, maple, basswood, elm, hickory, sycamore, beech, and tulip trees have all been used for nesting. Nests are constructed of bark, hair, and spiderwebs, among other materials. If a nest fails, the spiderweb from the old nest is often reused to build a new nest. The typical clutch consists of three to five creamy white eggs with brown speckling. When leaving the nest, the female "bungee jumps" with closed wings and drops toward the forest floor before spreading her wings and flying to her destination, likely in an effort to disguise the exact location of her nest.

Prairie Warbler
(*Setophaga discolor*)

CONSERVATION CONCERN SCORE: 13 (Moderate)

OTHER DESIGNATIONS: American Bird Conservancy watch list (Yellow),
2021 USFWS Birds of Conservation Concern, 2016 Partners in Flight Species of
Continental Concern (Yellow), 2021 Audubon Priority Birds list, Species of Greatest
Conservation Need (FL, KY, MD, MS, NC, PA, SC, TN, WV)

ESTIMATED POPULATION TREND 1966–2019: −65%

SIZE: Length 5 inches; wingspan 7 inches

Male Prairie Warblers exhibit yellow on the face and breast,
strong black facial markings, black streaks on the flanks,
weak wing bars, and chestnut streaks on the back.

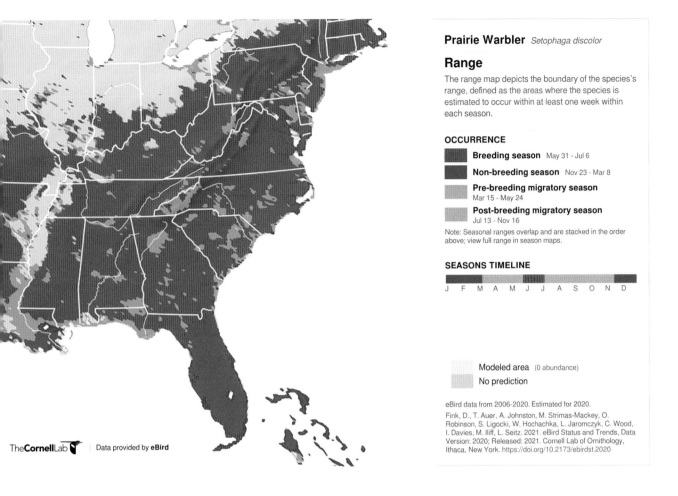

Prairie Warbler *Setophaga discolor*

Range

The range map depicts the boundary of the species's range, defined as the areas where the species is estimated to occur within at least one week within each season.

OCCURRENCE

Breeding season May 31 - Jul 6

Non-breeding season Nov 23 - Mar 8

Pre-breeding migratory season Mar 15 - May 24

Post-breeding migratory season Jul 13 - Nov 16

Note: Seasonal ranges overlap and are stacked in the order above; view full range in season maps.

SEASONS TIMELINE

J F M A M J J A S O N D

Modeled area (0 abundance)

No prediction

eBird data from 2006-2020. Estimated for 2020.

Fink, D., T. Auer, A. Johnston, M. Strimas-Mackey, O. Robinson, S. Ligocki, W. Hochachka, L. Jaromczyk, C. Wood, I. Davies, M. Iliff, L. Seitz. 2021. eBird Status and Trends, Data Version: 2020; Released: 2021. Cornell Lab of Ornithology, Ithaca, New York. https://doi.org/10.2173/ebirdst.2020

The**Cornell**Lab | Data provided by **eBird**

Species Account Prairie Warbler is a misleading name for this bird of shrublands, brush, and old fields, although admittedly, Scrub Warbler or Clear-Cut Warbler sounds less appealing. But one would be hard pressed to find this bird in a true open prairie habitat, as it prefers early successional wooded habitat often associated with poor soil. Overgrown orchards and Christmas tree farms, shrubby clear-cuts, areas recovering from fire, and old farm fields with patches of shrubs and young trees are all likely places to find this species. Disturbed places such as clear-cuts, burned areas, or abandoned fields are suitable for Prairie Warblers beginning about five years after the disturbance, and they remain in use for ten to twenty years. The birds seem to key in on areas with the right mix of open spaces, dense shrubs, and small trees, along with the absence of a closed canopy of mature trees.

With the logging of the eastern forests, the Prairie Warbler enjoyed a range expansion and likely reached peak numbers sometime in the 1940s and 1950s, when it was recorded nesting as far north as the Upper Peninsula of Michigan. According to Breeding Bird Survey data, however, the Prairie Warbler has suffered a significant decline throughout its range since the 1960s. Across the Appalachians and the Southeast, this decline generally ranges from 50 to 75 percent, although West Virginia has recorded an incredible 95 percent decline and Mississippi has actually seen a population increase. Overall, population losses are likely caused by changing forestry practices and fire suppression, which have resulted in a shift toward older forests. Habitat loss due to development and agriculture in the Caribbean, where this species winters, may be contributing to the decline.

Other factors negatively impacting the Prairie Warbler include nest parasitism by Brown-headed Cowbirds, forcing the warbler parents to raise cowbirds instead of their own chicks. In some cases, cowbird parasitism causes the warblers to abandon a nest altogether. In the northern parts of their breeding range, a second nesting attempt may not be possible if the first nest fails. Prairie Warblers also seem to suffer unusually high mortality rates, especially as fledglings. Only about 20 percent of Prairie Warbler eggs hatch and result in chicks that survive long enough to leave the nest. And of these successful fledglings, 79 percent perish within the first year of life, before they can return to the breeding grounds as adults. Natural predators include chipmunks, snakes, and other birds.

Identification Males are bright lemon yellow on the face, throat, breast, and under-parts, with bold black markings on the flanks. A black line runs through the eye, and a black malar stripe extends from the corner of the beak down the side of the throat. The male's back is olive green with chestnut streaks on the upper part. The female is similar, but with muted colors. There are two very faint wing bars. Prairie Warblers are "tail waggers," meaning that they habitually pump their tails up and down while perched. However, they seem to exhibit this behavior less frequently than Palm Warblers and Kirtland's Warblers. Similar species include the Magnolia Warbler (which has a darker back and much more white on the wings) and the Kirtland's Warbler (which has a slate-blue back).

Vocalizations The song of the Prairie Warbler includes up to ten buzzy *zee-zee-zee* notes that go up in pitch. Males sing two versions of this song that vary in volume and speed—one to attract females, and one to tell other males where the edges of the singing male's territory are located.

Nesting Some Prairie Warblers winter in Florida and may start to arrive in their breeding grounds elsewhere in the Southeast as early as mid to late March, although most arrivals take place in April. Males often return to the same territories year after year if the habitat is suitable. The nest consists of a cup made of plant fibers and lined with moss, grasses, hair, or feathers. The nest can be situated as high as forty-five feet but is often less than ten feet above the ground. According to a study of 608 Prairie Warbler nests, the most common tree species used for nesting is the American elm (Nolan 1978). Prairie Warblers lay three to five whitish eggs with brown speckles.

Female Prairie Warblers are more weakly marked than males.

In the fall, the Prairie Warbler's colors are slightly muted, but the distinctive black markings on the face and flanks remain.

This male Prairie Warbler is loudly proclaiming his territory,
apparently oblivious to the snail sharing his perch.

Canada Warbler
(*Cardellina canadensis*)

CONSERVATION CONCERN SCORE: 14 (High)

OTHER DESIGNATIONS: American Bird Conservancy watch list (Yellow),
2021 USFWS Birds of Conservation Concern, 2016 Partners in Flight Species of
Continental Concern (Yellow), 2021 Audubon Priority Birds list, Species of Greatest
Conservation Need (KY, MD, PA, TN, VA, WV)

ESTIMATED POPULATION TREND 1966–2019: −51%

SIZE: Length 5 inches; wingspan 8 inches

The male Canada Warbler sports a black necklace across a
yellow chest, black facial markings, and a slaty-blue back.

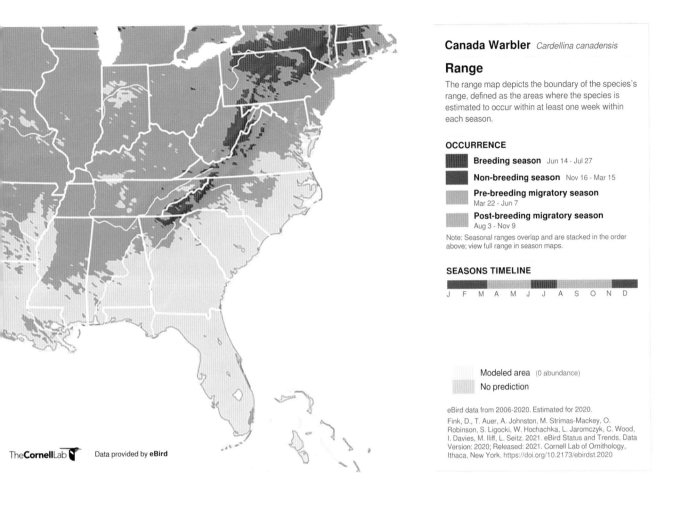

Canada Warbler *Cardellina canadensis*

Range

The range map depicts the boundary of the species's range, defined as the areas where the species is estimated to occur within at least one week within each season.

OCCURRENCE

Breeding season   Jun 14 - Jul 27

Non-breeding season   Nov 16 - Mar 15

Pre-breeding migratory season
Mar 22 - Jun 7

Post-breeding migratory season
Aug 3 - Nov 9

Note: Seasonal ranges overlap and are stacked in the order above; view full range in season maps.

SEASONS TIMELINE

J F M A M J J A S O N D

Modeled area   (0 abundance)

No prediction

eBird data from 2006-2020. Estimated for 2020.

Fink, D., T. Auer, A. Johnston, M. Strimas-Mackey, O. Robinson, S. Ligocki, W. Hochachka, L. Jaromczyk, C. Wood, I. Davies, M. Iliff, L. Seitz. 2021. eBird Status and Trends, Data Version: 2020; Released: 2021. Cornell Lab of Ornithology, Ithaca, New York. https://doi.org/10.2173/ebirdst.2020

TheCornellLab   Data provided by **eBird**

Species Account   Unlike other warbler species, the Canada Warbler doesn't spend all its time in the treetops. On their breeding grounds, Canada Warblers forage within fifteen feet of the ground. During migration, foraging heights typically range between five and thirty-five feet, with females staying slightly closer to the ground than males. Canada Warblers hover to glean prey from leaves and also catch insects on the wing—a habit that earned them their former name: Canadian Flycatcher. Many winged insects are taken, including mosquitoes, flies, and moths. Spiders and caterpillars are also consumed, as well as a few berries.

This bird is well named, as more than 80 percent of the Canada Warbler population nests in Canada, with the remainder nesting primarily in the northeastern United States. A small number of birds nest in the Appalachian Mountains from Pennsylvania south through West Virginia, Virginia, North Carolina, and northern Georgia. The birds occur in cool, moist forest with a dense understory and mossy ground cover for nesting. Canada Warblers may be locally abundant in young forest (six to thirty years after a disturbance caused by fire, timber harvesting, or storms). Interestingly, Canada

Warblers respond negatively to high deer populations, likely due to the resulting loss of understory vegetation. A study from Massachusetts found that in forests with one to three deer per square kilometer, eighty Canada Warblers were observed, whereas in similar forests with thirteen to twenty-three deer per square kilometer, only one Canada Warbler was observed.

The Canada Warbler is one of the later spring migrants, with early arrivals reaching the Appalachians in late April and the bulk arriving in May. It is also one of the first warblers to leave the breeding grounds in the fall. It is usually gone from the southeastern United States by September or early October as it makes its way to Panama and northern South America.

Breeding Bird Survey data indicate that the greatest decline in this species' breeding population is occurring in Canada, where the vast majority of these birds nest. It is estimated that the breeding population in Canada may have declined as much as 85 percent since 1968, with recent trends showing an acceleration in losses. In fact, according to Breeding Bird Survey data, as much as 43 percent of the breeding population in Canada was lost in the ten-year period from 1997 to 2007. As a result, this species is listed as threatened in Canada. Habitat loss, especially on wintering grounds in South America, is thought to be the primary reason for this decline. An estimated 95 percent of the cloud forest within the Canada Warbler's wintering range has been deforested since the 1970s. In Colombia alone, the rate of deforestation in the early 1990s was between 1.5 million and 2.2 million acres per year. This extreme habitat loss may well explain the steep decline of the Canada Warbler in recent decades. This species also seems especially susceptible to hitting obstructions while migrating at night. A remarkable 131 birds were killed in one night after colliding with a chimney in Ontario, Canada. In another case, twenty-seven Canada Warblers were killed when they collided with a television tower in Illinois.

Identification The Canada Warbler has slate-blue upperparts, lemon-yellow underparts, a whitish eye ring, a yellow line from the beak to the eye, and a black "necklace" of short, vertical streaks across its throat. Males also have some black on the cheeks and small black streaks on the crown. Females are similarly patterned but less strikingly marked. The wings and tail of both sexes are unmarked. Similar species include the Magnolia Warbler, which has a black or olive back instead of grayish blue and black streaking down the flanks, in addition to a similar black necklace. The Kirtland's Warbler is also similar but has black streaks on the back and white wing bars that the Canada Warbler lacks.

Vocalizations The song begins with a quick chip note, followed by a short burst of rich, warbled notes that sound rushed or hurried. The song lasts only about two seconds.

Nesting The female builds the nest either on the ground or very close to it, usually in sphagnum hummocks or other mossy areas. Canada Warblers lay three to five creamy white eggs speckled with brown. Incubation lasts approximately twelve days, and chicks remain in the nest for an additional ten days before fledging. Both parents care for the nestlings and feed them insects.

The female Canada Warbler is duller than the male but also has a faint black necklace.

Scarlet Tanager
(*Piranga olivacea*)

CONSERVATION CONCERN SCORE: 11 (Moderate)

OTHER DESIGNATIONS: 2021 USFWS Birds of Conservation Concern, 2021 Audubon Priority Birds list, Species of Greatest Conservation Need (MD, MS, PA, SC)

ESTIMATED POPULATION TREND 1966–2019: –5%

SIZE: Length 7 inches; wingspan 12 inches

The male Scarlet Tanager is a brilliant, almost fluorescent red, with jet-black wings and tail.

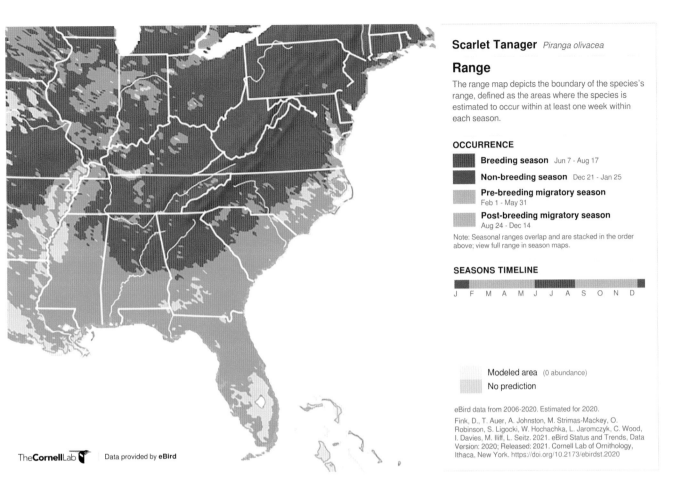

Scarlet Tanager *Piranga olivacea*

Range

The range map depicts the boundary of the species's range, defined as the areas where the species is estimated to occur within at least one week within each season.

OCCURRENCE

	Breeding season	Jun 7 - Aug 17
	Non-breeding season	Dec 21 - Jan 25
	Pre-breeding migratory season	Feb 1 - May 31
	Post-breeding migratory season	Aug 24 - Dec 14

Note: Seasonal ranges overlap and are stacked in the order above; view full range in season maps.

SEASONS TIMELINE

J F M A M J J A S O N D

Modeled area (0 abundance)

No prediction

eBird data from 2006-2020. Estimated for 2020.

Fink, D., T. Auer, A. Johnston, M. Strimas-Mackey, O. Robinson, S. Ligocki, W. Hochachka, L. Jaromczyk, C. Wood, I. Davies, M. Iliff, L. Seitz. 2021. eBird Status and Trends, Data Version: 2020; Released: 2021. Cornell Lab of Ornithology, Ithaca, New York. https://doi.org/10.2173/ebirdst.2020

TheCornellLab   Data provided by **eBird**

Species Account Spotting a male Scarlet Tanager in the forests of the Appalachians is like finding a palm tree growing in a snowbank. With their fiery red plumage, tanagers seem too exotic and tropical for the eastern United States. Strong fliers, Scarlet Tanagers do, in fact, spend much of the year in the tropics. They winter along the eastern slopes of the Andes Mountains in South America from Colombia south through portions of Ecuador, Brazil, and Peru. In spring, they migrate north through Central America and across the Gulf of Mexico, arriving in April along the Gulf coast. The exhausted migrants often feed in trees and shrubs at or below eye level, providing ample opportunity to spy these stunning birds as they rest and recover lost energy reserves before resuming their journey. Upon arriving on their breeding grounds, they often stay high in the canopy of oak woods, making observation much more difficult.

Scarlet Tanagers need a specific type of habitat to breed successfully. Studies such as Cornell University's Project Tanager have shown that tanagers have the highest reproductive success when they nest in large, unfragmented forest blocks within a larger landscape that is at least 70 percent forested. Both the size of the forest block itself and the percentage of forest within the surrounding 2,500 acres are important. If 70 percent of the surrounding landscape is forested, tanagers can be highly successful in unfragmented forest blocks as small as 66 acres. However, if only 40 percent of the surrounding landscape is forested, an unfragmented forest block of at least 605 acres is required to achieve the same level of nesting success. The reason is that tanagers respond negatively to increasing amounts of forest edge habitat, which is created when forests are cleared for roads, houses, agricultural fields, or other land uses. These edge habitats allow species such as the Brown-headed Cowbird to invade the forest. Normally reluctant to enter unbroken forest, cowbirds take advantage of grassy openings around the forest edge, where they can feed and gain access to other birds' nests in the interior and lay their own eggs there. Forest edge and suburban habitats also lead to more feral cats and other predators that reduce tanagers' breeding success as well.

Scarlet Tanagers feed heavily on insects they glean from leaves or branches or catch in the air. Major prey species include cicadas, moths, butterflies, caterpillars, dragonflies, and beetles. Tanagers also frequently take bees, wasps, and other stinging insects, which they apparently consume without removing the stinger. Tanagers also consume a variety of fruits, including blackberries, serviceberries, raspberries, mulberries, and strawberries. Shortly after arriving on the breeding ground in the spring, while it is still cold, tanagers sometimes forage in leaf litter on the ground for earthworms and beetles when other insect prey is not readily available in the canopy.

As a whole, Scarlet Tanager numbers are relatively stable, with only slight declines recorded. However, it is a species of concern because of its need for large blocks of forest. Pairs that nest in fragmented forest or small forest patches may actually be population sinks—meaning that these birds produce nearly zero young and therefore contribute to overall population declines. Significant declines have been noted regionally, specifically in the Northeast and upper Midwest, likely due to forest fragmentation.

Identification The male Scarlet Tanager is difficult to misidentify, with his flaming red head, back, and belly and jet-black wings and tail. The male Summer Tanager is a less brilliant red, and its wings and tail are red rather than black. Females of the two species are very similar. The female Scarlet Tanager is greenish yellow, while the female Summer Tanager is more lemon yellow. The female Scarlet Tanager also has more black in the wing. Both species have heavy, light-colored beaks. Young male tanagers sport yellow plumage with splotches of bright red during their first year of life.

Vocalizations Scarlet Tanagers have a hoarse, raspy, warbling song similar to that of an American Robin, but slower and less clear. Both sexes sing. The common call is a *tick-burr,* which is often the first clue to the bird's presence high in the canopy.

Nesting Nests are built by the female in the canopy of mature forest, often fifty feet or more off the ground. The nests are rather loosely constructed over the course of a few days and consist of roots, grasses, twigs, bark, and other material. Three to five bluish eggs, speckled with brown and purplish markings, are laid.

Female Scarlet Tanagers are olive-yellow with black wings, making it hard to believe they are the same species as their flame-colored mates.

During migration, Scarlet Tanagers often forage low in the canopy or even in the shrub layer, making this a good time of year to observe the species. During the breeding season, they often stay in the canopy of mature forests.

Painted Bunting
(*Passerina ciris*)

CONSERVATION CONCERN SCORE: 11 (Moderate)

OTHER DESIGNATIONS: American Bird Conservancy watch list (Yellow), 2021 USFWS Birds of Conservation Concern, State Protected (AL), State Special Concern (GA, NC), Species of Greatest Conservation Need (FL, SC, TN)

ESTIMATED POPULATION TREND 1966–2019: −35%

SIZE: Length 6 inches; wingspan 9 inches

Male Painted Buntings are a remarkable palette of neon colors, reflected here in the water as the bird bathes in a shallow pool.

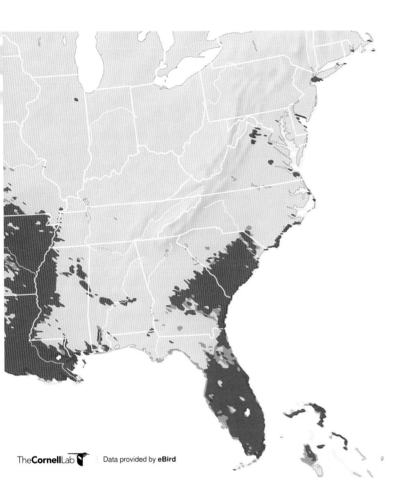

eBird data from 2006-2020. Estimated for 2020.
Fink, D., T. Auer, A. Johnston, M. Strimas-Mackey, O. Robinson, S. Ligocki, W. Hochachka, L. Jaromczyk, C. Wood, I. Davies, M. Iliff, L. Seitz. 2021. eBird Status and Trends, Data Version: 2020; Released: 2021. Cornell Lab of Ornithology, Ithaca, New York. https://doi.org/10.2173/ebirdst.2020

Species Account Like a vibrant rainbow come to life, the male Painted Bunting is one of the most strikingly beautiful birds in North America. With a purple-blue head, scarlet belly and rump, and lime-green back, the male Painted Bunting seems more likely to reside in the tropical forests of the Amazon than alongside sparrows in grassy fields, hedgerows, and weedy vacant lots in the southeastern United States. It takes males two years to develop their adult plumage. Females and first-year males are beautiful in their own right, with green backs and lemony-yellow bellies. The species is also known as the Nonpareil, meaning "unrivaled" or "unparalleled." Anyone who has thumbed through the pages of a field guide can attest to this.

There is a large gap in the species' breeding range that essentially splits the population into two distinct groups. Historically, they have been treated as subspecies. The birds that breed along the southern Atlantic coast from North Carolina to Florida are smaller and darker than the population that breeds from Mississippi west to Texas and extreme southeastern New Mexico and north to southern Kansas and Missouri. The Atlantic coast population winters in southern Florida and the Caribbean, while the south-central population winters in Mexico and Central America. The timing of the feather molt differs between the two populations, and some genetic distinctions have been documented as well, leading some scientists to believe that the two populations are actually separate species.

Although somewhat secretive and typically preferring to stay fairly low to the ground, males defending a breeding territory sing their bubbling song from the foliage near the treetops. Territorial battles between males can be fierce and may involve beating each other with their wings, pecking, and grappling with their feet. Serious wounds can be inflicted during these battles, including the loss of an eye or even death. In a study conducted in Georgia, male Painted Buntings established territories roughly five acres in size. Higher-quality territories tended to be located along woodland edges, where prey was more abundant. During a two-year period, 95 percent of the males retained the same territory in both breeding seasons (Lanyon and Thompson 1986). Males that can successfully defend high-quality territories may mate with multiple females.

Although the decline of the Painted Bunting population since the 1960s is concerning, much of this drop took place in the 1960s and 1970s. Since 1980, numbers seem to have stabilized, although the Atlantic coast population remains especially vulnerable because of ongoing coastal development and its relatively small breeding area, which totals only about 4 percent of the more western population's breeding grounds. In addition to habitat loss, capture of wild Painted Buntings for the pet trade continues to threaten the species' future. As early as the 1800s, Painted Buntings were captured and shipped from New Orleans to markets in Europe. Although this is now illegal in the United States, the capture and sale of Painted Buntings continue in parts of Mexico, Cuba, and elsewhere. It is estimated that at least 100,000 Painted Buntings were captured for the pet trade in Mexico between 1982 and 1999.

Identification With its purple-blue head, red eye ring, brilliant scarlet underparts and rump, and lime-green back, the adult male Painted Bunting is unmistakable. Young males and adult females are also distinctive, with bluish-green upperparts and a yellow-green chest and belly.

Vocalizations Male Painted Buntings give a rich, liquid, warbling song, usually from an elevated perch. The call is a short, metallic *tsick*.

Nesting Most nests are placed less than ten feet off the ground and are built exclusively by the female. The cup-shaped nest is often lined with fine grass or hair and contains a typical clutch of three or four white or pale gray eggs speckled with brown at the larger end.

Female Painted Buntings are predominantly
a grass-green color, with shades of teal
and aqua on the upper wing. Young birds
resemble the female but are much grayer.

Painted Buntings often feed close to the ground, where they search for seeds and insects near fencerows, thickets, or other scrubby habitats.

Dickcissel
(*Spiza americana*)

CONSERVATION CONCERN SCORE: 11 (Moderate)

OTHER DESIGNATIONS: 2021 USFWS Birds of Conservation Concern,
Species of Greatest Conservation Need (KY, MD, PA, SC, WV)

ESTIMATED POPULATION TREND 1966–2019: −28%

SIZE: Length 6 inches; wingspan 10 inches

Male Dickcissels are sparrow-sized birds with a black
and yellow pattern on the chest, similar to a meadowlark.
The red shoulder patch is a good field mark.

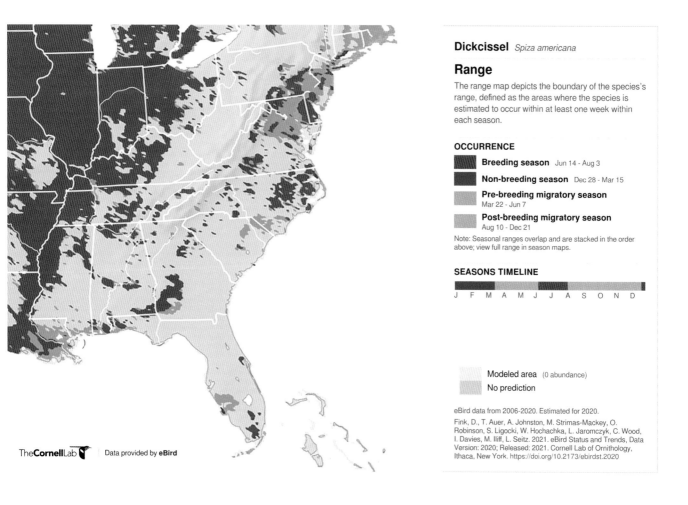

eBird data from 2006-2020. Estimated for 2020.

Fink, D., T. Auer, A. Johnston, M. Strimas-Mackey, O. Robinson, S. Ligocki, W. Hochachka, L. Jaromczyk, C. Wood, I. Davies, M. Iliff, L. Seitz. 2021. eBird Status and Trends, Data Version: 2020; Released: 2021. Cornell Lab of Ornithology, Ithaca, New York. https://doi.org/10.2173/ebirdst.2020

Species Account The Dickcissel was once abundant throughout the Great Plains and stretched into the grassland habitats of the Southeast all the way to the Atlantic coastline in the 1800s. Today, the birds are still found in small numbers in parts of their former range, especially in Kentucky, Tennessee, and Mississippi. This contraction of much of the Dickcissel's eastern range might be related to a declining overall population, or it could be connected to changes in land use.

Male Dickcissels are persistent vocalizers, starting their rhythmic, chattering song and singing incessantly almost as soon as they arrive on their breeding territory, which is usually about a week prior to the arrival of the females. Dickcissels are considered polygynous, but whether males breed with more than one female depends on the quality of the territory they control. Territories with dense grasses and forbs, providing many suitable nesting locations, tend to be more attractive to females. A male whose territory contains poor nesting habitat may fail to attract a mate. In a study of 150 male Dickcissels, 14 percent failed to attract a mate, 36 percent had one mate, 41 percent had two mates, and 9 percent had three or more mates (Zimmerman 1966).

Dickcissels are considered an agricultural pest across much of their wintering range. Although some birds winter in Mexico and Central America, the bulk of the wintering population is found in the llanos region of Venezuela. This area was once dominated by grasslands and wetlands but has now been largely converted to agriculture. Rice and sorghum farmers there consider the huge flocks of Dickcissels that feed and roost in their grain fields during the nonbreeding season a threat to their livelihoods. Some have resorted to shooting or clubbing the birds, driving through their roosts to destroy them, or even illegally poisoning large numbers of Dickcissels by having crop dusters spray nighttime roosts with toxic organophosphates, leading to massive die-offs. It is possible that some of the largest roosts in Venezuela contain 10 to 30 percent of the world's Dickcissel population, so targeting these areas could have significant consequences. Despite the drastic actions taken by farmers, a 1992 study revealed that Dickcissels actually consumed only about 1 percent of the total rice and sorghum crops in Venezuela.

In addition to this persecution on their wintering grounds, Dickcissels have lost significant breeding habitat in North America. Only a fraction of the original prairie remains in the core nesting area of the Great Plains, and hayfields and grassland habitats have declined across the Southeast over the last century. Because of this combination of stressors, some ornithologists predicted in the late 1970s that the Dickcissel would be extinct by 2000. Fortunately, the Conservation Reserve Program created millions of acres of grassland habitat in the United States, and the Dickcissel and many other declining grassland species benefited from this improvement. Mass poisonings in Venezuela have decreased, although extermination programs continue at the local level, despite laws protecting the birds. Population trends since the early 1980s have apparently stabilized, although some declines continue.

Identification Male Dickcissels are gray overall, with rufous on the shoulder and wing, brown and black on the back, and a black bib on a yellow chest. Females are similar but not as brightly colored.

Vocalizations The male sings frequently, with two or three short introductory notes followed by a harsh, buzzy jumble of notes. The bird's name originates from its song, which sounds like *dic-dic cissel*.

Nesting The nest is built on the ground, and the female lays four pale blue, unmarked eggs. The hatchlings are fed only by the female and fledge from the nest within a week to ten days.

This female Dickcissel is perched on the stem of a prairie dock plant and carries a caterpillar, apparently to feed a clutch of chicks nearby.

ACKNOWLEDGMENTS

I am deeply thankful to my family for their ongoing support, encouragement, and understanding through the long hours needed to complete a project like this. I'm also grateful to many who contributed comments and expertise that strengthened this work. However, I am most grateful to the Lord Jesus Christ. It is His incredible creation that, in some small way, this book seeks to appreciate and celebrate, for "through him all things were made; without him nothing was made that has been made" (John 1:3). As we study creation, we learn more about the Creator who designed everything with incredible imagination, balance, beauty, and wisdom. He has charged us with stewarding this creation, and I pray that He gives us the wisdom to do that well. May He be glorified in a small way by the pages of this book.

REFERENCES

Alabama Department of Conservation and Natural Resources. 2015. "Alabama's
Wildlife Action Plan." https://www.outdooralabama.com/sites/default/files
/Research/SWCS/AL_SWAP_FINAL% 20June2017.pdf. Accessed 30 August
2022.

Allen, R. P. 1952. *The Whooping Crane*. Report 3. National Audubon Society.

American Bird Conservancy. 2020. "Swallow-tailed Kite Forest Management Recommendations for Forest Owners in the Southeastern US." https://abcbirds.org
/wp-content/uploads/2020/05/STKI-Fact-Sheet_Spring-2020.pdf. Accessed 3
April 2021.

American Bird Conservancy. n.d. "ABC's Bird Library—Gull-billed Tern." https://
abcbirds.org/bird/gull-billed-tern/. Accessed 7 August 2021.

American Bird Conservancy. n.d. "ABC's Bird Library—King Rail." https://abcbirds
.org/bird/king-rail/. Accessed 30 April 2022.

American Bird Conservancy. n.d. "United States Watchlist of Birds of Conserva-
tion Concern." https://www.nps.gov/goga/learn/management/upload/-1929
-American-Bird-Conservancy_National-Audubon-Society_2007_DONE.pdf.
Accessed 5 May 2020.

American Oystercatcher Working Group, E. Nol, and R. C. Humphrey. 2020. "American Oystercatcher (*Haematopus palliatus*)," version 1.0. In *Birds of the World*,
edited by A. F. Poole. Ithaca, NY: Cornell Lab of Ornithology. https://doi.org/10
.2173/bow.ameoys.01.

Anich, N. M., T. J. Benson, J. D. Brown, C. Roa, J. C. Bednarz, R. E. Brown, and J. G.
Dickson. 2020. "Swainson's Warbler (*Limnothlypis swainsonii*)," version 1.0. In
Birds of the World, edited by P. G. Rodewald. Ithaca, NY: Cornell Lab of Ornithol-
ogy. https://doi.org/10.2173/bow.swawar.01.

Arnold, J. M., S. A. Oswald, I. C. T. Nisbet, P. Pyle, and M. A. Patten. 2020.
"Common Tern (*Sterna hirundo*)," version 1.0. In *Birds of the World*, edited by
S. M. Billerman. Ithaca, NY: Cornell Lab of Ornithology. https://doi.org/10.2173
/bow.comter.01.

Audubon, J. J. 1827. *Birds of America*. London: Havell.

Audubon Florida. 2012. "Status Brief on the Endangered Florida Grasshopper Sparrow." https://fl.audubon.org/sites/default/files/audubon_grasshopper sparrow_statusupdate_july2012_1.pdf. Accessed 28 August 2021.

Audubon Florida. n.d. "Fantastic News for Florida Grasshopper Sparrows." https://fl.audubon.org/news/fantastic-news-florida-grasshopper-sparrows. Accessed 28 August 2021.

Audubon Florida. n.d. "Florida Grasshopper Sparrow." https://fl.audubon.org/birds/florida-grasshopper-sparrow. Accessed 28 August 2021.

Axelson, G. n.d. "Even Small, Scattered Florida Scrub-Jay Groups Are Vital to the Survival of the Species." Cornell Lab of Ornithology, All About Birds. https://www.allaboutbirds.org/news/even-small-scattered-florida-scrub-jay-groups-are -vital-to-the-survival-of-the-species/. Accessed 25 January 2022.

Baker, A., P. Gonzalez, R. I. G. Morrison, and B. A. Harrington. 2020. "Red Knot (*Calidris canutus*)," version 1.0. In *Birds of the World*, edited by S. M. Billerman. Ithaca, NY: Cornell Lab of Ornithology. https://doi.org/10.2173/bow.redkno.01.

Bent, A. C. (1929) 1962. "The Wilson's Plover." In *Life Histories of North American Shore Birds*. New York: Dover Publications.

Bentley, E. L. 1994. "Use of a Landscape-Level Approach to Determine the Habitat Requirements of the Yellow-crowned Night-Heron, *Nycticorax violaceus*, in the Lower Chesapeake Bay." Master's thesis, College of William and Mary.

Bielefeld, R. R., M. G. Brasher, T. E. Moorman, and P. N. Gray. 2020. "Mottled Duck (*Anas fulvigula*)," version 1.0. In *Birds of the World*, edited by A. F. Poole. Ithaca, NY: Cornell Lab of Ornithology. https://doi.org/10.2173/bow.motduc.01.

Bohlen, H. D. 1989. *The Birds of Illinois*. Bloomington: Indiana University Press.

Brennan, L. A., F. Hernandez, and D. Williford. 2020. "Northern Bobwhite (*Colinus virginianus*)," version 1.0. In *Birds of the World*, edited by A. F. Poole. Ithaca, NY: Cornell Lab of Ornithology. https://doi.org/10.2173/bow.norbob.01.

Brown, S., C. Hickey, B. Harrington, and R. Gill, eds. 2001. *The U.S. Shorebird Conservation Plan*. 2nd ed. Manomet, MA: Manomet Center for Conservation Sciences.

Bryan, D. C. 1981. "Territoriality and Pair Bonding in the Limpkin (*Aramus guarauna*)." Master's thesis, Florida State University.

Bryan, D. C. 2020. "Limpkin (*Aramus guarauna*)," version 1.0. In *Birds of the World*, edited by A. F. Poole and F. B. Gill. Ithaca, NY: Cornell Lab of Ornithology. https://doi.org/10.2173/bow.limpki.01.

Buehler, D. A., P. B. Hamel, and T. Boves. 2020. "Cerulean Warbler (*Setophaga cerulea*)," version 1.0. In *Birds of the World*, edited by A. F. Poole. Ithaca, NY: Cornell Lab of Ornithology. https://doi.org/10.2173/bow.cerwar.01.

Canadian Wildlife Service and US Fish and Wildlife Service. 2005. *International Recovery Plan for the Whooping Crane*. Ottawa: Recovery of Nationally Endangered Wildlife (RENEW); Albuquerque, NM: USFWS.

Canterbury, R. A., Jr., N. J. Kotesovec, and B. Catuzza. 1995. "A Preliminary Study of the Effects of Brown-headed Cowbird Parasitism on the Reproductive Success of Blue-winged Warblers in Northeastern Ohio." *Ohio Cardinal* 18:124–25.

Chesser, R. T., S. M. Billerman, K. J. Burns, C. Cicero, J. L. Dunn, A. W. Kratter, I. J. Lovette, N. A. Mason, P. C. Rasmussen, J. V. Remsen Jr., D. F. Stotz, and K. Winker. 2020. "Check-list of North American Birds." American Ornithological Society. https://checklist.americanornithology.org/taxa/.

Confer, J. L., P. Hartman, and A. Roth. 2020. "Golden-winged Warbler (*Vermivora chrysoptera*)," version 1.0. In *Birds of the World*, edited by A. F. Poole. Ithaca, NY: Cornell Lab of Ornithology. https://doi.org/10.2173/bow.gowwar.01.

Confer, J. L., and J. L. Larkin. 2003. "Effects of Vegetation, Interspecific Competition, and Brood Parasitism on Golden-winged Warbler (*Vermivora chrysoptera*) Nesting Success." *Auk* 120(1):138–44.

Cornell Lab of Ornithology. n.d. "All About Birds." https://www.allaboutbirds.org/. Accessed 2021.

Coulter, M. C., J. A. Rodgers Jr., J. C. Ogden, and F. C. Depkin. 2020. "Wood Stork (*Mycteria americana*)," version 1.0. In *Birds of the World*, edited by A. F. Poole and F. B. Gill. Ithaca, NY: Cornell Lab of Ornithology. https://doi.org/10.2173/bow.woosto.01.

Dunning, J. B., Jr., P. Pyle, and M. A. Patten. 2020. "Bachman's Sparrow (*Peucaea aestivalis*)," version 1.0. In *Birds of the World*, edited by P. G. Rodewald. Ithaca, NY: Cornell Lab of Ornithology. https://doi.org/10.2173/bow.bacspa.01.

Elliott-Smith, E., and S. M. Haig. 2020. "Piping Plover (*Charadrius melodus*)," version 1.0. In *Birds of the World*, edited by A. F. Poole. Ithaca, NY: Cornell Lab of Ornithology. https://doi.org/10.2173/bow.pipplo.01.

Evans, M., E. Gow, R. R. Roth, M. S. Johnson, and T. J. Underwood. 2020. "Wood Thrush (*Hylocichla mustelina*)," version 1.0. In *Birds of the World*, edited by A. F. Poole. Ithaca, NY: Cornell Lab of Ornithology. https://doi.org/10.2173/bow.woothr.01.

Fink, D., T. Auer, A. Johnston, M. Strimas-Mackey, S. Ligocki, O. Robinson, W. Hochachka, L. Jaromczyk, A. Rodewald, C. Wood, I. Davies, and A. Spencer. 2022. *eBird Status and Trends, Data Version: 2021*. Ithaca, NY: Cornell Lab of Ornithology. https://doi.org/10.2173/ebirdst.2021.

Florida Fish and Wildlife Conservation Commission. 2019. "Florida State Wildlife Action Plan." https://myfwc.com/media/22767/2019-action-plan.pdf. Accessed 30 August 2022.

Forbush, E. H. 1912. *A History of Game Birds, Wild Fowl, and Shore Birds of Massachusetts and Adjacent States.* 2nd ed. Boston: Massachusetts State Board of Agriculture.

Frei, B., K. G. Smith, J. H. Withgott, P. G. Rodewald, P. Pyle, and M. A. Patten. 2020. "Red-headed Woodpecker (*Melanerpes erythrocephalus*)," version 1.0. In *Birds of the World*, edited by P. G. Rodewald. Ithaca, NY: Cornell Lab of Ornithology. https://doi.org/10.2173/bow.rehwoo.01.

Georgia Department of Natural Resources, Wildlife Resources Division. 2015. "Georgia State Wildlife Action Plan." https://georgiawildlife.com/sites/default /files/wrd/pdf/swap/appendix-a-high-priority-species-and-habitats-summary -data.pdf. Accessed 30 August 2022.

Gill, F. B., R. A. Canterbury, and J. L. Confer. 2020. "Blue-winged Warbler (*Vermivora cyanoptera*)," version 1.0. In *Birds of the World*, edited by A. F. Poole and F. B. Gill. Ithaca, NY: Cornell Lab of Ornithology. https://doi.org/10.2173/bow .buwwar.01.

Gochfeld, M., J. Burger, and K. L. Lefevre. 2020. "Black Skimmer (*Rynchops niger*)," version 1.0. In *Birds of the World*, edited by S. M. Billerman. Ithaca, NY: Cornell Lab of Ornithology. https://doi.org/10.2173/bow.blkski.01.

Greenlaw, J. S., C. S. Elphick, W. Post, and J. D. Rising. 2020. "Saltmarsh Sparrow (*Ammospiza caudacuta*)," version 1.0. In *Birds of the World*, edited by P. G. Rodewald. Ithaca, NY: Cornell Lab of Ornithology. https://doi.org/10.2173/bow .sstspa.01.

Herkert, J. R., P. D. Vickery, and D. E. Kroodsma. 2020. "Henslow's Sparrow (*Centronyx henslowii*)," version 1.0. In *Birds of the World*, edited by P. G. Rodewald. Ithaca, NY: Cornell Lab of Ornithology. https://doi.org/10.2173/bow.henspa.01.

Hoover, J. P., and M. C. Brittingham. 1993. "Regional Variation in Cowbird Parasitism of Wood Thrushes." *Wilson Bulletin* 105:228–38.

Hurdle, J. 2021. "DNREC Says New Report of Big Decline in Shorebirds Is Only a 'Snapshot.'" Delaware Public Media. https://www.delawarepublic.org/science -health-tech/2021-06-18/dnrec-says-new-report-of-big-decline-in-rare-shorebird -is-only-a-snapshot. Accessed 23 July 2021.

Jackson, J. A. 2020. "Red-cockaded Woodpecker (*Dryobates borealis*)," version 1.0. In *Birds of the World*, edited by A. F. Poole and F. B. Gill. Ithaca, NY: Cornell Lab of Ornithology. https://doi.org/10.2173/bow.recwoo.01.

Jehl, J. R., Jr., J. Klima, and R. E. Harris. 2020. "Short-billed Dowitcher (*Limnodromus griseus*)," version 1.0. In *Birds of the World*, edited by A. F. Poole and F. B. Gill. Ithaca, NY: Cornell Lab of Ornithology. https://doi.org/10.2173/bow.shbdow.01.

Journey North. n.d. "Whooping Crane." https://journeynorth.org/tm/crane /Population.html. Accessed 5 March 2022.

Kaufman, K. n.d. "Clapper Rail." National Audubon Society Field Guide. https://www.audubon.org/field-guide/bird/clapper-rail. Accessed 29 June 2023.

Kentucky Department of Fish and Wildlife Resources. 2013. "Kentucky State Wildlife Action Plan." https://fw.ky.gov/WAP/Pages/Default.aspx. Accessed 30 August 2022.

Klinger, S. 2011. *Northern Bobwhite Quail Management Plan for Pennsylvania*. Pennsylvania Game Commission.

Koczur, L. M., M. C. Green, B. M. Ballard, P. E. Lowther, and R. T. Paul. 2020. "Reddish Egret (*Egretta rufescens*)," version 1.0. In *Birds of the World*, edited by P. G. Rodewald. Ithaca, NY: Cornell Lab of Ornithology. https://doi.org/10.2173/bow.redegr.01.

Kozicky, E. L., and F. W. Schmidt. 1949. "Nesting Habits of the Clapper Rail in New Jersey." *Auk* 66:355–64.

Kurose, S. 2021. "Interior Least Tern Is Latest Endangered Species Act Success Story." Center for Biological Diversity. https://biologicaldiversity.org/w/news/press-releases/interior-least-tern-latest-endangered-species-act-success-2021-01-12/. Accessed 2 May 2022.

Lanyon, S. M., and C. F. Thompson. 1986. "Site Fidelity and Habitat Quality as Determinants of Settlement Pattern in Male Painted Buntings." *Condor* 88:206–10.

Loss, S. R., T. Will, S. S. Loss, and P. Marra. 2014. "Bird–Building Collisions in the United States: Estimates of Annual Mortality and Species Vulnerability." *Condor* 116(1):8–23.

Loss, S. R., T. Will, and P. Marra. 2013. "The Impact of Free-Ranging Domestic Cats on Wildlife of the United States." *Nature Communications* 4:1396.

Lowther, P. E., H. D. Douglas III, and C. L. Gratto-Trevor. 2020. "Willet (*Tringa semipalmata*)," version 1.0. In *Birds of the World*, edited by A. F. Poole and F. B. Gill. Ithaca, NY: Cornell Lab of Ornithology. https://doi.org/10.2173/bow.willet1.01.

Lowther, P. E., S. M. Lanyon, C. W. Thompson, and T. S. Schulenberg. 2020. "Painted Bunting (*Passerina ciris*)," version 1.0. In *Birds of the World*, edited by S. M. Billerman. Ithaca, NY: Cornell Lab of Ornithology. https://doi.org/10.2173/bow.paibun.01.

Maryland Department of Natural Resources. 2015. "Maryland State Wildlife Action Plan." https://dnr.maryland.gov/wildlife/Pages/plants_wildlife/SWAP_Submission.aspx. Accessed 30 August 2022.

McDonald, M. V. 2020. "Kentucky Warbler (*Geothlypis formosa*)," version 1.0. In *Birds of the World*, edited by A. F. Poole. Ithaca, NY: Cornell Lab of Ornithology. https://doi.org/10.2173/bow.kenwar.01.

McKay, W. D. 1981. "Notes on Purple Gallinules in Colombian Rice Fields." *Wilson Bulletin* 93:267–71.

Meyer, K. D. 2020. "Swallow-tailed Kite (*Elanoides forficatus*)," version 1.0. In *Birds of the World*, edited by A. F. Poole and F. B. Gill. Ithaca, NY: Cornell Lab of Ornithology. https://doi.org/10.2173/bow.swtkit.01.

Meyer, K. D., and M. W. Collopy. 1990. *Status, Distribution, and Habitat Requirements of the American Swallow-tailed Kite* (Elanoides forficatus forficatus) *in Florida*. Tallahassee: Florida Game and Fresh Water Fish Commission.

Mississippi Department of Wildlife, Fisheries, and Parks. 2015. "State Wildlife Action Plan." https://www.mdwfp.com/media/251788/mississippi_swap_revised_16_september_2016__reduced_pdf. Accessed 30 August 2022.

Molina, K. C., R. M. Erwin, E. Palacios, E. Mellink, and N. W. H. Seto. 2010. *Status Review and Conservation Recommendations for the Gull-billed Tern* (Gelochelidon nilotica) *in North America*. Biological Technical Publication FWS/BTP-R1013-2010. Washington, DC: US Department of the Interior, USFWS.

Mordecai, R. S. 2008. *Chimney Watch: Providing a Foundation for Coordinated Monitoring of Urban Aerial Insectivores*. Northeast Coordinated Bird Monitoring Partnership and American Bird Conservancy.

Mowbray, T. B. 2020. "Scarlet Tanager (*Piranga olivacea*)," version 1.0. In *Birds of the World*, edited by A. F. Poole and F. B. Gill. Ithaca, NY: Cornell Lab of Ornithology. https://doi.org/10.2173/bow.scatan.01.

National Park Service. n.d. "Wood Stork: Species Profile." https://www.nps.gov/ever/learn/nature/woodstork.htm. Accessed 28 August 2021.

Nisbet, I. C. T., C. S. Mostello, R. R. Veit, J. W. Fox, and V. Afanasyev. 2011. "Migrations and Winter Quarters of Five Common Terns Tracked Using Geolocators." *Waterbirds* 34(1):32–39.

Nolan, V., Jr. 1978. *The Ecology and Behavior of the Prairie Warbler* Dendroica discolor. Ornithological Monograph 26. Washington, DC: American Ornithologists' Union.

Nolan, V., Jr., E. D. Ketterson, and C. A. Buerkle. 2020. "Prairie Warbler (*Setophaga discolor*)," version 1.0. In *Birds of the World*, edited by A. F. Poole. Ithaca, NY: Cornell Lab of Ornithology. https://doi.org/10.2173/bow.prawar.01.

North American Bird Conservation Initiative, State of North America's Birds. 2016. "Species Assessment Summary and Watchlist." https://www.stateofthebirds.org/2016/resources/species-assessments/.

North Carolina Wildlife Resources Commission. 2015. "North Carolina Wildlife Action Plan." http://www.ncwildlife.org/plan.aspx. Accessed 30 August 2022.

Oney, J. 1951. "Fall Food Habits of the Clapper Rail in Georgia." *Journal of Wildlife Management* 15:106–7.

Owl Pages. n.d. "Burrowing Owl—*Athene cunicularia*." https://www.owlpages.com/owls/species.php?s=2250. Accessed 28 August 2021.

Owl Research Institute. n.d. "Burrowing Owl—*Athene cunicularia*." https://www .owlresearchinstitute.org/burrowing-owl. Accessed 28 August 2021.

Page, G. W., L. E. Stenzel, J. S. Warriner, J. C. Warriner, and P. W. Paton. 2020. "Snowy Plover (*Charadrius nivosus*)," version 1.0. In *Birds of the World*, edited by A. F. Poole. Ithaca, NY: Cornell Lab of Ornithology. https://doi.org/10.2173/bow .snoplo5.01.

Palmer, R. S. 1949. "Maine Birds." *Bulletin of the Museum of Comparative Zoology* (Harvard University) 102:1–579.

Partners in Flight. 2016. "Species of Continental Concern." https://www .partnersinflight.org/wp-content/uploads/2016/07/SPECIES-OF-CONT -CONCERN-from-pif-continental-plan-final-spread-2.pdf. Accessed 10 February 2020.

Pennsylvania Fish and Boat Commission. 2015. "Pennsylvania Wildlife Action Plan." https://www.fishandboat.com/Resource/Documents/SWAP-CHAPTER -1-apx13.pdf. Accessed 30 August 2022.

Petit, L. J. 2020. "Prothonotary Warbler (*Protonotaria citrea*)," version 1.0. In *Birds of the World*, edited by A. F. Poole and F. B. Gill. Ithaca, NY: Cornell Lab of Ornithology. https://doi.org/10.2173/bow.prowar.01.

Pickens, B. A., and B. Meanley. 2020. "King Rail (*Rallus elegans*)," version 1.0. In *Birds of the World*, edited by P. G. Rodewald. Ithaca, NY: Cornell Lab of Ornithology. https://doi.org/10.2173/bow.kinrai4.01.

Post, W., and J. S. Greenlaw. 2020. "Seaside Sparrow (*Ammospiza maritima*)," version 1.0. In *Birds of the World*, edited by P. G. Rodewald. Ithaca, NY: Cornell Lab of Ornithology. https://doi.org/10.2173/bow.seaspa.01.

Poulin, R. G., L. D. Todd, E. A. Haug, B. A. Millsap, and M. S. Martell. 2020. "Burrowing Owl (*Athene cunicularia*)," version 1.0. In *Birds of the World*, edited by A. F. Poole. Ithaca, NY: Cornell Lab of Ornithology. https://doi.org/10.2173/bow .burow1.01.

Powell, H. 2008. "The Cornell Lab All About Birds." https://www.allaboutbirds.org /news/scrubland-survivor-the-florida-scrub-jay/. Accessed 25 January 2022.

Reichert, B. E., C. E. Cattau, R. J. Fletcher Jr., P. W. Sykes Jr., J. A. Rodgers Jr., and R. E. Bennetts. 2020. "Snail Kite (*Rostrhamus sociabilis*)," version 1.0. In *Birds of the World*, edited by A. F. Poole. Ithaca, NY: Cornell Lab of Ornithology. https:// doi.org/10.2173/bow.snakit.01.

Reitsma, L. R., M. T. Hallworth, M. McMahon, and C. J. Conway. 2020. "Canada Warbler (*Cardellina canadensis*)," version 2.0. In *Birds of the World*, edited by P. G. Rodewald and B. K. Keeney. Ithaca, NY: Cornell Lab of Ornithology. https://doi .org/10.2173/bow.canwar.02.

Riegner, M. F. 1982. "The Diet of Yellow-crowned Night-Herons in the Eastern and Southern United States." *Colonial Waterbirds* 5:173–76.

Rising, J. 2005. "Ecological and Genetic Diversity in the Seaside Sparrow." *Birding*, September–October, 490–96.

Rodgers, J. A., Jr. 1983. "Foraging Behavior of Seven Species of Herons in Tampa Bay, Florida." *Colonial Waterbirds* 6:11–23.

Rodgers, J. A., Jr., and H. T. Smith. 2020. "Little Blue Heron (*Egretta caerulea*)," version 1.0. In *Birds of the World*, edited by A. F. Poole. Ithaca, NY: Cornell Lab of Ornithology. https://doi.org/10.2173/bow.libher.01.

Rosenberg, K. V., R. W. Rohrbaugh Jr., S. E. Barker, J. D. Lowe, R. S. Hames, and A. A. Dhondt. 1999. *A Land Managers Guide to Improving Habitat for Scarlet Tanagers and Other Forest-Interior Birds*. Ithaca, NY: Cornell Lab of Ornithology.

Rosenberg, K. V., et al. 2019. "Decline of North American Avifauna." *Science* 366(6461):120–24.

Rush, S. A., K. F. Gaines, W. R. Eddleman, and C. J. Conway. 2020. "Clapper Rail (*Rallus crepitans*)," version 1.0. In *Birds of the World*, edited by P. G. Rodewald. Ithaca, NY: Cornell Lab of Ornithology. https://doi.org/10.2173/bow.clarai1.01.

Rush, S. A., J. A. Olin, A. T. Fisk, M. S. Woodrey, and R. J. Cooper. 2010. "Trophic Relationships of a Marsh Bird Differ between Gulf Coast Estuaries." *Estuaries and Coasts* 33(4):963–70.

Saha, P. 2015. "Piping Plovers Get a Protected Park in the Bahamas." Audubon. https://www.audubon.org/magazine/november-december-2015/piping-plovers-get-protected-park. Accessed 11 December 2021.

Savage, A. L., C. E. Moorman, J. A. Gerwin, and C. Sorenson. 2010. "Prey Selection by Swainson's Warblers on the Breeding Grounds." *Condor* 112:605–14.

Schulte, S., S. Brown, D. Reynolds, and the American Oystercatcher Working Group. 2010. *American Oystercatcher Conservation Action Plan for the United States Atlantic and Gulf Coasts*. Version 2.1. http://www.conservewildlifenj.org/downloads/cwnj_310.pdf.

Slater, G. L., J. D. Lloyd, J. H. Withgott, and K. G. Smith. 2021. "Brown-headed Nuthatch (*Sitta pusilla*)," version 1.1. In *Birds of the World*, edited by A. F. Poole. Ithaca, NY: Cornell Lab of Ornithology. https://doi.org/10.2173/bow.bnhnut.01.1.

Smallwood, J. A., and D. M. Bird. 2020. "American Kestrel (*Falco sparverius*)," version 1.0. In *Birds of the World*, edited by A. F. Poole and F. B. Gill. Ithaca, NY: Cornell Lab of Ornithology. https://doi.org/10.2173/bow.amekes.01.

Smith, K. G., S. R. Wittenberg, R. B. Macwhirter, and K. L. Bildstein. 2020. "Northern Harrier (*Circus hudsonius*)," version 1.0. In *Birds of the World*, edited by P. G. Rodewald. Ithaca, NY: Cornell Lab of Ornithology. https://doi.org/10.2173/bow.norhar2.01.

Soots, R. F., Jr., and J. F. Parnell. 1975. "Ecological Succession of Breeding Birds in Relation to Plant Succession on Dredge Islands in North Carolina." University of North Carolina Sea Grant Program.

Sousa, B. F., S. A. Temple, and G. D. Basili. 2022. "Dickcissel (*Spiza americana*)," version 2.0. In *Birds of the World*, edited by T. S. Schulenberg and B. K. Keeney. Ithaca, NY: Cornell Lab of Ornithology. https://doi.org/10.2173/bow.dickci.02.

South Carolina Department of Natural Resources. 2015. "South Carolina State Wildlife Action Plan." https://www.dnr.sc.gov/swap/main/2015StateWildlifeActionPlan-chaptersonly.pdf. Accessed 30 August 2022.

Spahn, S. A., and T. W. Sherry. 1999. "Cadmium and Lead Exposure Associated with Reduced Growth Rates, Poorer Fledging Success of Little Blue Heron Chicks (*Egretta caerulea*) in South Louisiana Wetlands." *Archives of Environmental Contamination and Toxicology* 37(3):377–84.

Sperry, C. C. 1940. *Food Habits of a Group of Shorebirds: Woodcock, Snipe, Knot, and Dowitcher*. Washington, DC: US Government Printing Office.

Steeves, T. K., S. B. Kearney-McGee, M. A. Rubega, C. L. Cink, and C. T. Collins. 2020. "Chimney Swift (*Chaetura pelagica*)," version 1.0. In *Birds of the World*, edited by A. F. Poole. Ithaca, NY: Cornell Lab of Ornithology. https://doi.org/10.2173/bow.chiswi.01.

Summerour, B. 2008. "Summary of Breeding Data for the Swainson's Warbler (*Limnothlypis swainsonii*) in Alabama, 1912–2004." *Alabama Birdlife* 54:1–4.

Tennessee Wildlife Resources Agency. 2015. "Tennessee State Wildlife Action Plan." http://twraonline.org/2015swap.pdf. Accessed 30 August 2022.

Tennessee Wildlife Resources Agency. n.d. "Prothonotary Warbler Nest Box." https://www.tn.gov/twra/wildlife/woodworking-for-wildlife/prothonotary-warbler-nest-box.html. Accessed 6 May 2022.

Thompson, B. C., J. A. Jackson, J. Burger, L. A. Hill, E. M. Kirsch, and J. L. Atwood. 2020. "Least Tern (*Sternula antillarum*)," version 1.0. In *Birds of the World*, edited by A. F. Poole and F. B. Gill. Ithaca, NY: Cornell Lab of Ornithology. https://doi.org/10.2173/bow.leater1.01.

Thoreau, H. D. 1853. "The Wood Thrush." Personal journal.

Three Billion Birds. n.d. "3 Billion Birds Gone." https://www.3billionbirds.org/. Accessed 25 May 2022.

Urbanek, R. P., and J. C. Lewis. 2020. "Whooping Crane (*Grus americana*)," version 1.0. In *Birds of the World*, edited by A. F. Poole. Ithaca, NY: Cornell Lab of Ornithology. https://doi.org/10.2173/bow.whocra.01.

US Fish and Wildlife Service. 1991. *Mississippi Sandhill Crane Recovery Plan*. Atlanta: USFWS.

US Fish and Wildlife Service. 1996. *Piping Plover* (Charadrius melodus), *Atlantic Coast Population, Revised Recovery Plan*. Hadley, MA: USFWS.

US Fish and Wildlife Service. 2021. *Birds of Conservation Concern 2021*. Falls Church, VA: US Department of the Interior, USFWS.

US Fish and Wildlife Service. n.d. "*Calidris canutus rufu.*" https://fws.gov/northeast
/red-knot/. Accessed 25 May 2022.

US Fish and Wildlife Service. n.d. "FWS-Listed Species by Taxonomic Group-
Birds." Environmental Conservation Online System. https://ecos.fws.gov/ecp
/report/species-listings-by-tax-group?statusCategory=Listed&groupName=
Birds. Accessed 25 May 2022.

US Fish and Wildlife Service. n.d. "Mississippi Sandhill Crane National Wildlife
Refuge." https://www.fws.gov/refuge/mississippi. Accessed 2 February 2022.

US Fish and Wildlife Service. n.d. "Proposed Downlisting of the Red-cockaded
Woodpecker from Endangered to Threatened." https://www.fws.gov/southeast
/faq/proposed-downlisting-of-the-red-cockaded-woodpecker-from-endangered
-to-threatened/. Accessed 14 August 2021.

Virginia Department of Wildlife Resources. 2015. "2015 Virginia Wildlife Action
Plan." http://bewildvirginia.org/wildlife-action plan/pdf/Final% 20SGCN%
20List% 20Appendix% 20A% 20July% 202016.pdf. Accessed 30 August 2022.

Watts, B. D. 2000. *Bachman's Sparrow Management Plan: Fort A. P. Hill, Virginia.*
Center for Conservation Biology Technical Report Series CCBTR-00-06. Wil-
liamsburg, VA: College of William and Mary.

Watts, B. D. 2020. "Yellow-crowned Night-Heron (*Nyctanassa violacea*)," version 1.0.
In *Birds of the World*, edited by A. F. Poole. Ithaca, NY: Cornell Lab of Ornithology. https://doi.org/10.2173/bow.ycnher.01.

West, R. L., and G. K. Hess. 2020. "Purple Gallinule (*Porphyrio martinica*)," version
1.0. In *Birds of the World*, edited by A. F. Poole and F. B. Gill. Ithaca, NY: Cornell
Lab of Ornithology. https://doi.org/10.2173/bow.purga12.01.

West Virginia Division of Natural Resources. n.d. "West Virginia State Wildlife
Action Plan." https://wvdnr.gov/wp-content/uploads/2021/05/2015-West
-Virginia-State-Wildlife-Action-Plan-Submittal-1.pdf. Accessed 30 August 2022.

Weston, F. M. 1968. "Bachman's Sparrow." In "Life Histories of North American
Cardinals, Grosbeaks, Buntings and Allies," edited by O. L. Austin. *Bulletin of the
United States National Museum* 237:956–75.

White Oak Conservation. n.d. "Florida Grasshopper Sparrow." https://www
.whiteoakwildlife.org/wildlife/florida-grasshopper-sparrow/. Accessed 28 August
2021.

Will, T. C. 1986. "The Behavioral Ecology of Species Replacement: Blue-winged and
Golden-winged Warblers in Michigan." PhD diss., University of Michigan–Ann
Arbor.

Williams, M. 2018. *Endangered and Disappearing Birds of the Midwest.* Bloomington:
Indiana University Press.

Wilson, A. 1831. *The Natural History of the Birds of the United States.* Edinburgh:
Constable.

Wojtowicz, S. 2016. "The Search Is on for Piping Plovers." US Fish and Wildlife Service Northeast Region. https://usfwsnortheast.wordpress.com/2016/03/09/the-search-is-on-for-piping-plovers/. Accessed 12 January 2022.

Woolfenden, G. E., and J. W. Fitzpatrick. 2020. "Florida Scrub-Jay (*Aphelocoma coerulescens*)," version 1.0. In *Birds of the World*, edited by A. F. Poole and F. B. Gill. Ithaca, NY: Cornell Lab of Ornithology. https://doi.org/10.2173/bow.flsjay.01.

Yosef, R. 2020. "Loggerhead Shrike (*Lanius ludovicianus*)," version 1.0. In *Birds of the World*, edited by A. F. Poole and F. B. Gil. Ithaca, NY: Cornell Lab of Ornithology. https://doi.org/10.2173/bow.logshr.01.

Zdravkovic, M. G. 2013. *Conservation Plan for the Wilson's Plover* (Charadrius wilsonia). Version 1.0. Manomet, MA: Manomet Center for Conservation Sciences.

Zdravkovic, M. G., C. A. Corbat, and P. W. Bergstrom. 2020. "Wilson's Plover (*Charadrius wilsonia*)," version 1.0. In *Birds of the World*, edited by P. G. Rodewald. Ithaca, NY: Cornell Lab of Ornithology. https://doi.org/10.2173/bow.wilplo.01.

Zimmerman, J. L. 1966. "Polygyny in the Dickcissel." *Auk* 83:534–46.

ABOUT THE AUTHOR

An award-winning nature photographer and author, Matt Williams is the current director of Conservation Programs for the Indiana chapter of the Nature Conservancy. His career with the organization began in 1998 as a seasonal field assistant banding endangered birds and mapping their nesting habitat on the Fort Hood military base in central Texas, part of a joint project with the Department of Defense. From there, he became the preserve manager for the Nature Conservancy's Texas City Prairie Preserve, where he led habitat restoration efforts and worked closely with the Attwater's Prairie Chicken Recovery Team to track the population of this critically endangered species.

Matt came to work for the Indiana chapter in 2001 as the North-Central Indiana land steward. He became a burn boss and led prescribed fire and invasive species control efforts across more than a dozen Nature Conservancy properties in that part of the state. In his current role, he oversees the science, land acquisition, and land management work of the chapter. He also is the past chair and a current member of the Midwest Division Conservation Cabinet, which is responsible for developing and reviewing many of the conservation strategies being implemented across the five states of the Midwest Division.

Matt's books include *Indiana State Parks: A Centennial Celebration,* which features his landscape photography, and *Endangered and Disappearing Birds of the Midwest,* which tells the story of forty bird species across the Midwest that are most in need of conservation efforts. The latter won a silver medal in the 2018 Foreword Reviews Independent Book Awards competition and a bronze medal in the 2019 Independent Publisher Book Awards—a competition involving books from nearly 2,500 publishers from the United States, Canada, England, Australia, and other countries. He wrote the foreword for a new printing of the classic work *The Birds of Indiana,* featuring the life histories of many of the bird species of the state as well as the artwork of William Zimmerman. Matt's photographs have appeared in several other books and national magazines, including *Bird Watcher's Digest* and *National Wildlife.*

Matt, his wife Karyn, and their four children live in Crawfordsville, Indiana, where they enjoy kayaking along Sugar Creek, hiking at Shades State Park, and serving in their local church.